INSIDE THE SECRET SPACE PROGRAMS

Herbert Dorsey

outskirts
press

Inside the Secret Space Programs
All Rights Reserved.
Copyright © 2017 Herbert Dorsey
v4.0

The opinions expressed in this manuscript are solely the opinions of the author and do not represent the opinions or thoughts of the publisher. The author has represented and warranted full ownership and/or legal right to publish all the materials in this book.

This book may not be reproduced, transmitted, or stored in whole or in part by any means, including graphic, electronic, or mechanical without the express written consent of the publisher except in the case of brief quotations embodied in critical articles and reviews.

Outskirts Press, Inc.
http://www.outskirtspress.com

Paperback ISBN: 978-1-4787-8347-3

Outskirts Press and the "OP" logo are trademarks belonging to Outskirts Press, Inc.

PRINTED IN THE UNITED STATES OF AMERICA

Table of Contents

The Early History of Airships 1
Secret Space Programs 11
Other Science 30
NASA 49
Hyper Dimensional Physics 72
Recovered ET Space Craft 82
Technology for the Secret Navy Space Fleet 106
Colonizing the Solar System 149
Current Events on Earth 170
Conclusions 188
Other Books by this Author: 195
Bibliography 196

Preface

The governments of the world have been quite secretive about UFOs, extraterrestrial interactions and extraterrestrial treaties with their governments (or exopolitics) and their secret space programs. And, the United States of America is undoubtedly the worst offender.

This secrecy has primarily served two purposes.

One, to keep calm among the world's populations and make them believe nothing unusual is going on although thousands of sightings of UFOs have been seen all over the planet. The powers that be fear that their hold over mankind, through religion, economics, and a controlled news media would be destabilized if mankind knew the truth about our interplanetary and interstellar brothers and sisters, most who come to our planet to help us spiritually evolve and to prevent man's senseless destruction of each other and the planet we live on.

Two, to keep suppressed the advanced technology either received directly from the extraterrestrials or back engineered from retrieved crashed UFOs. This technology would free mankind from economic bondage imposed by the need for energy, shelter, transportation, pharmaceuticals, expensive surgery, and other vital needs, like cleaning up our over polluted environment.

Also, some of this technology secretly held by secret organizations have the ability to make them seem like magicians or even gods. Try to imagine the advantages one who could travel in time, or teleport through space to other planets or other locations on this planet, might have over ordinary humans. Try to imagine the many ways this type of technology could be misused.

I have already touched on these subjects in two previous books; *Secret Science and the Secret Space Program,* and *The Covert Colonization of Our Solar System.* However, since writing the last book, new whistle blowers and inside informants, like William Tompkins, Clifford Stone, Tony Rodrigues and others who have come forward and revealed what they knew about the secret space programs, underground military bases, exotic technology, and crashed UFO retrievals.

William Tompkins, a man who played a large role in NASA's Apollo program, designing the launch facility building and working out the logistics of the program, also played a large role in a secret program, designing large space ships for the U.S. Navy. These space ships were huge multi kilometer long ships that use antigravity technology and have interstellar travel abilities. They were built in a large, secret underground facility built in a large cavern located north of Salt Lake City, Utah, as revealed in interviews with Tomkins on the Rense Radio show and in his book, *Selected by Extraterrestrials.*

Much of the technology used in the secret space program comes from recovered crashed UFOs and captured Nazi flying saucers as well as Nazi scientists brought to the U.S. in Operation Paperclip.

However, even more important than the hardware recovered and back engineered from crashed UFOs are the occupants of these unfortunate craft, that manage to survive.

This is because they make it possible to gain information on things like; what is their mission, how does their technology work, who are the other ETs visiting our planet, and what are their agendas etc. That is where Clifford Stone and others with telepathic ability come in as "interfacers" with these crashed UFO survivors to learn more about them and their civilizations.

For many years Clifford Stone kept his work secret, even from his family, as his employers wished. However, after the death of his son, he decided that his family had a right to know why he would disappear for great lengths of time on these covert operations. He also decided that the tax payers that were paying for these programs also had a right to know what was going on. And so, Clifford Stone has gone public with his formerly top secret work.

Tony Rodrigues relates his experiences with the Illuminati and the Nazi *Dark Fleet,* which was so secretive that even Corey Goode knew little about it.

Also, I cover a little history here. There seems to be even older secret societies that were behind the great air ship mysteries [not lighter than air] of the late 1890s, which is explored in greater detail here.

Additionally, "other science" leading to free energy, new healing technologies, antigravity, transmutation, and a non-secret space program that is now totally out in the open, run by the Keshe Foundation Space Ship Institute will be looked at.

From these and a number of other sources, I have learned enough new information for another interesting book. To some it will appear to be science fiction. But, I assure you that, as far as I could ascertain with my considerable research into the matter- it is not!

So, here it is. *Inside the Secret Space Programs*. Learn and Enjoy!

The Early History of Airships

The airships I will be dealing here are not the lighter than air variety that are well known about.

There are another variety of airships, the working principles of which are little known about or understood. They were developed by individual inventors and secret societies in the U.S. starting just before the Civil War and by the 1890s had created a number of working airships.

In Europe, there were also individual groups experimenting with heavier than air airships that worked on entirely different principals than displacement of air by less dense gasses like hot air, hydrogen or helium.

One theory, put forth by Walter Bosley, in his book *Origin:The 19th Century Emergence of the 20th Century Breakaway Civilizations* is as follows:

> There existed ancient civilizations that possessed technology far in advanced of our present time. Proof exists in ancient texts like Hindu Mahabharata that speak of flying Vimana's. Secret societies have passed some the

secrets of these extinct civilizations along up to the present time. Also, societies have launched expeditions to ancient sites to uncover or rediscover these secrets.

The technology of these 19th Century airship builders was based on lost technology of lost civilizations preserved and rediscovered. And, the 20th Century Nazi Germany technology linked to the modern secret space program originated with the Prussian airship builders of the 19th Century.

Bosley's theory is reinforced by a translator for the Vatican, Mauro Biglino, who translated many ancient Hebrew scriptures for the Vatican. The only problem was that the exact translations of these Old Testament parchments contradicted much of the official Catholic doctrine. So, finally Biglino was fired from his job.

After that, he wrote a number of books in Italian on what is wrong about the accepted Bible. One book that was translated into English is titled, *The Book That Will Forever Change Our Ideas About The Bible*. The reason this book concerns us is that it greatly reinforces the ancient astronaut theories developed by researchers like Zechariah Sitchin and Erich von Daniken.

These ancient extraterrestrial astronauts (the Elohim/ Anunnaki) created humans in their own image by genetic manipulations on ape like primates using part of their own DNA

Biglino's translations show that "angels" were of flesh and blood, but were of the extraterrestrial races that were technically and intellectually far in advance of humankind. The God of the Jews, Yahweh, was one of these "angles" who was very jealous of the other "Gods" while himself being quite mortal.

One would have to ask themselves why would "God" tell Israelites to make war on Canaan, if he wasn't jealous of the "Gods" of the Canaanites. It seems that the Elohim or the Anunnaki had their own different groups that fought among themselves, and used humans as their servants and armies while demanding their worship, as revealed by Zachariah Sitchin. Do these sound like wise and spiritually evolved beings?

This Judaic "God" definitely was not at one with the true source of all that is and is definitely not a "God" that I would love, respect or worship.

The God I worship has love for all its creations and does not wish his children make war on each other. War is a perversion of spirituality and is the destructive domain of Satan. Jesus, who was at one with the true nameless God which he called "the father", and who was not the military "messiah" the Jews were hoping for, advised his apostles to be as wise as serpents and as harmless as doves.

Later, the Catholic Church of Rome, would greatly pervert the actual teachings of Christ to use for their own geopolitical agendas, as revealed in my *The Secret History of the New World Order* and other histories.

In any case, along with other advanced technology, the Elohim/Anunnaki possessed their own space ships and flying devices. And these secret technologies were passed along by the scientists/priests, which in those times were considered the same, in the mystery schools of the ancient Illuminati. This is the lost technology Walter Bosley was speaking of.

Walter Bosley is an interesting person himself. He worked for the Air Force Office of Special Investigations (AFOSI), a U.S. federal law enforcement agency that reports directly to the Office of the Secretary of the Air Force. Operating worldwide, AFOSI provides independent criminal investigative, counterintelligence and protective service operations outside of the traditional military chain of command. So, his abilities at investigative research should not be underestimated.

Also, there existed two well-known experimenters in the U.S. that worked on aircraft that were neither lighter than air nor worked on aerodynamic principals, James Worrell Keely and Nikola Tesla, as discussed in my *Secret Science and the Secret Space Program*.

According to researcher Jerome Clark, from the 1880s to the 1890s, numerous airship sightings occurred around the world. Some of these airships had the appearance of dirigibles with a gas bag supporting the passenger compartment below and powered with propellers. However, the size of the gas bags was much too small for sufficient lift of the weight suspended below, even using hydrogen gas!

A Secret Society called the Aero Club existed in Sonora, California. In 1858, it was headed by Peter Mennis and seemed to be headed by an even more secret organization in Prussia (not New York, as I mistakenly stated in a previous book) with the initials NYMZA. The Sonora Aero Club had a number of airships with gas bags supporting the passenger compartment below. However, none of these gas bags, using hydrogen, could possibly displace enough air to lift the weight of the passenger compartment. It was claimed that a special NB gas was used that possessed special antigravity properties.

The primary source of information on the Aero Club is Charles Dellschau, an immigrant from Prussia, who became a member of the Aero Club in Sonora.

He revealed, in his diaries, that many of the airships, the members called "aeros", were actually built by a secretive group of Prussian immigrants in Tuolumne County, California. He said that at that time, the Aero Club was led by Peter Mennis, who knew the secret recipe for the fuel used for propulsion and lift mechanism used in their craft. These craft were flying from the mid to late 1850s!

Dellschau also was an artist and created drawings and paintings of these "aeros." These pictures can be seen in Walter Bosley's book *Origin: The 19th Century Emergence of the 20th Century Breakaway Civilizations*. An examination of these pictures clearly indicate that these airships were definitely not lighter than air craft.

Dellschau had asked Peter Mennis how the craft worked. This is the explanation he received:

> The aeros possess two chambers that the engine mechanism fills with NB gas, produced from a secret blend of chemicals and water.
>
> This blend was applied to a rotating drum via a series of hoses and valves and the spinning process created the NB gas which in turn spins the shaft with which the pilot controls the direction of the aeros. This NB gas was the key as it provided the propulsion and lift for the craft.

According to Dellschau, the first flight of Peter Mennis' aero occurred on a winter day in 1857 in a field outside of Columbia, California (which is now an airport). The aero ascended to 100 feet and maneuvered back and forth across the field all day.

Other copies of this aero soon followed, created by the club members.

Soon, some members of the Aero Club were considering developing these aeros for commercial and military purposes. However, most of the members, including Peter Mennis were opposed.

Nevertheless, soon a Prussian military officer, probably from NYMZA in Prussia, approached the Aero Club and suggested that their aeros be designed for warfare. The Aero Club was mostly opposed to the idea as was Peter Mennis, and the Prussian military officer was sent away empty handed.

Not long after this incidence, Peter Mennis was badly burned in an aero accident and died. The secret mixture to create NB gas was never revealed by Peter Mennis to the other members and so, the aeros stopped working after his death. This led to the dissolution of the Aero Club by 1860.

However, according to Dellschau there was one member of the Aero Club, Tosh Wilson who had developed a huge aero he called the *Aero EE* that could lift a payload of 30,000 pounds which was of a size far beyond the aeros pictured by Dellschau. This advanced design seemed more like the mystery air ships that turned up in 1896 and 1897.

It could be possible that Tosh Wilson had discovered the secret formula that created NB gas and also kept it secret from the other Members of the Aero Club and later continued on his own.

There is the distinct possibility that NYMZA in Prussia, the parent society behind the Aero Club, may have directed secret sabotage of Peter Mennis' aero after being turned down by their leader. This could even have been done by a member of the club with more loyalty to NYMZA than Peter Mennis.

According to Walter Bosley, NYMZA in German is really NJMZa since the J is pronounced Yah in German and NJMZa stands for Nationalistische Jagdflugzeug Maschinen Zahlungsamt which in English is: Nationalist Pursuit-Exploration Airship Program Office.

Another air ship was developed and patented by Solomon Andrews during the Civil War which he called the Aeron. This ship was definitely not lighter than air. It consisted of three 80-foot-long, cigar shaped, horizontal gas bags made of varnished linen and connected side by side, which supported a basket 12 feet long, containing 3 passengers and the drive equipment. Once again, the dimensions of the gas bags were too small to support the weight involved even if filled with hydrogen gas!

In a test flight for the Union Army, the Aeron could maintain a speed of 120 MPH and performed maneuvers for all to see. The motor, propulsion methods and secrets of lift were kept secret for fear the Confederate Army could discover the technology. This technology seems to still be secret.

Then, there were the great airship sightings of the 1890s. As reported in the *Dallas Morning News* and *Galveston News* on April 19, 1897 an airship was sighted on the ground at Stephenville, Texas by 22 persons, including a senator, judges, a mayor, the district attorney and reporters. They met with the airship pilot, S.E. Tillman and engineer, A.E. Dolbear who said that they were on an experimental expedition for their financial backers in New York. The airship was sixty feet long and cigar shaped with large wheels on both ends and was propelled by an electric engine powered by batteries.

It was quite likely that the pilot and engineer were Colonel Samuel Escue Tillman and Amos E. Dolbear.

Tillman was a career officer in the U.S. Army who taught at West point for 30 years, and was an engineer and astronomer. As a founding member of the Cosmos Club he was an associate of the men who created the National Geographic Society. He also wrote many books and articles. Notable topics of his writings included "Mixture of Gas and Vapor", Refraction of Heat", Steam Engines", Cyclones", and "Movements of High Pressure Areas". These topics could have a resemblance to creating the NB gas through rotational movement of certain chemicals. Also Tillman was an astronomer on the national expedition to Tasmania to observe the transit of Venus.

So, he was highly qualified to be a pilot and navigator on this experimental expedition. The Engineer, Amos E. Dolbear was also quite interesting.

Amos E. Dolbear earned his Ph.D. from the University of Michigan where he then became a professor of chemistry. He also was an inventor who perfected the electric gyroscope, developed a magneto electric telephone and demonstrated wireless communication in 1882, way before Tesla, Marconi and Hertz.

Both men were the sort of men that New York investors could trust and would financially back.

Both Nikola Tesla and Keely developed antigravity devices in the 19th century. As discussed *Secret Science and the Secret Space Program,* Keely's air ship was worked on and developed between 1888 and 1893 and eventually flew as was mentioned in the press. He later actually had his airship demonstrated to the U.S. War Department, which certified it in 1896. What ever became of it afterwards remains a mystery.

Tesla's electromagnetic flying machine was never patented. Probably because Tesla thought that mankind wasn't spiritually developed sufficiently to use the technology safely. He was occasionally seen near his Colorado Spring laboratory riding his electromagnetic flying machine around by some nearby ranchers.

Walter Bosley in his book *Origin: The 19th Century Emergence of the 20 th Century Breakaway Civilizations* is able to link the Prussian NYZMA or NJZMa to the creation of the German's Die Glocke or Nazi Bell. Which among other things was also an antigravity device.

Another interesting person is William Tomkins, who because of his photographic memory, was placed into Naval Intelligence at North Island, San Diego, right after completing boot camp in 1942. There, he was involved with debriefing 29 secret U.S. agents placed within the Nazi flying saucer programs. Tompkins would later become involved with designing antigravity space ships for the Navy.

As William Tompkins reveals in his book *Selected by Extraterrestrials*, Naval Intelligence discovered, before and during World War II, that there were 1,442 people in Europe attempting to build antigravity space ships capable of leaving the planet. His claim was that they were being telepathically inspired by Nordic extraterrestrials to build these space ships to help mankind to start exploring the Universe.

During the War, the Nazi SS arrested these people and forced them to work on the German secret antigravity programs. Near the end of the war however, they were released by the SS and collectively succeeded in building a large space ship and actually left the planet!

In conclusion, it seems that a variety of secret technologies were being used at the end of the 19th Century to create mystery air ships which seemed to use antigravity and were sighted in numerous locations on numerous occasions, but primarily in 1896 and 1897.

Some of this secret technology was carried over into early 20th century and became part of Nazi Germany's flying saucer program.

Secret Space Programs

The U.S. secret space program got its start during World War II. The German secret space program got started two decades earlier. German secret societies, The Vrill Gesellschaft, and the Thule Gesellschaft and a more secret society inside of Thule, Schwarze Sonne or Black Sun., working with the psychic, Maria Orsic, and professor, Dr. Winfried Otto Schumann, were developing a flying disk they called the Jenseitsflugmaschine (JFM, or in English "other world flying machine") as early as 1923. The first model, JFM 1 failed. But, the problems were corrected and they had a working model (JFM 2) by 1924.

The Vrill Gesellschaft or Vrill Society was originally comprised of female psychics or mediums that could channel information from spirit sources or beings on other star systems. The Thule Society was founded August 17, 1918 by Rudolf von Sebottendorff, a German occultist, as the Munich branch of the Germanenorden, a secret society also known as the "Order of Teutons". The secretive Black Sun society also played a role in the creation of Thule.

Maria Orsic, who founded the Vrill Society, was the primary channel for the information to build the JFM saucer shaped flying machine, which came from "the Sumi" on a planet orbiting a star they called Aldebaran in the Taurus Star system. These Sumi claimed that they started the Sumerian civilization on Earth in ancient times. This information was contained in automatic writing and also from pictures placed in Maria's mind which she later drew. The writing was in in an ancient Templar script and ancient Sumerian text which had to be translated into German by experts in ancient languages.

Maria Orsic could not understand the technical side of the translated information channeled. She asked her father, Tomislav Orsic, if he could understand it. After studying it, he said that he didn't understand it either. But, he would take the information to Professor Winfried Otto Schumann, who he knew at the University of Munich, and see if he could understand it.

The operational theory had much in common with the theories and work of both Victor Schauberger and Dr. Karl Schappeller, which Dr. Schumann was familiar with. Schauberger had learned how to tap the power of vortexes to create levitation. And Schappeller had developed a theory on the primary state of matter and experimentally created glowing magnetism, which was 1,000 times stronger than regular magnetism. The Schappeller device not only was able to create free energy but also levitated. Schumann was quite interested in Maria Orsic's information which had some similarities to Schappeller and Shauberger's theories and agreed to try to develop a working model.

Later, Schumann would become known in Germany as the father of the Vrill levitation technology. Money to develop these exotic technologies primarily came from the Thule Society.

These secret societies kept their flying saucers secret. It wasn't until the emergency of World War II that eventually the Nazi SS gained control of this technology. Under Hans Kammler's direction, this technology was weaponized and kept secret from the German people and the World at large. (1)

As another development of the Third Reich, Hitler had negotiated a treaty with a group of reptilian extraterrestrials, known as the Dracos.

In return for forming an alliance with the Dracos, the Nazis would receive from the Dracos working flying saucers, advanced weaponry and the rights to create a base on the Earth's moon.

Hitler met with the Draco reptilian leader in an underground cave. Hitler referred to the Draco reptilian leader as the "superman". Hitler claimed that "his eyes were terrible" and "I was afraid". This Draco "superman" was also directing Hitler in his efforts to conquer the world, which in a 1933 speech Hitler spoke of as creating the "New World Order".

The Nazis also had an alliance with the more peaceful Agarthans, who lived in cities within large caverns within the Earth's crust. The Agarthans had told the Nazi Anerbe explorers in Tibet about the location of one of their abandoned cities under the Antarctic which could be reached by their U-Boats in a region the Germans called Nue Schabenland. The Nazis located this abandoned underground city and rebuilt it, using slave labor, and called it Nue Berlin. (2)

During World War II, operatives from a secret U.S. Department, had penetrated some of the German secret flying saucer programs. These operatives were quite fluent in the German language and customs and otherwise were quite intelligent, crafty and good at their job. All of this information, including documents and photographs, was being transferred back to Naval Admiral, Rico Botta in the U.S. at North Island, San Diego, in a highly classified program, higher than Naval intelligence.

One person receiving this intelligence was William Mills Tompkins, who seven decades later, in 2015, started revealing what he knew of these secret war time programs, to the world, in his book, *Selected by Extraterrestrials*.

Because of his hobby of creating highly detailed scale Navy ship models from his excellent photographic memory, Tompkins, was selected by Lieutenant J.G. Perry Wood of Naval Intelligence to be assigned to Naval Intelligence, after Tompkins completed boot camp in 1942.

Tompkins was working under U.S. Navy Admiral, Rico Botta, who covertly was in charge of intelligence operations out of the Naval Air Station at North beach, San Diego, while serving as a Naval Aircraft maintenance officer. Also, Tomkins worked with three Navy captains who were specialists in different areas of Naval aircraft design, engineering, and procurement.

In addition to Admiral Rico Botta official daytime naval duty as "Assembly and Repair Officer" at Naval Air Station, San Diego, in the evenings he led a covert Navy program operating with 29 spies embedded in Nazi Germany's most advanced aerospace programs which were developing antigravity flying saucers. This covert program learned from the Navy operatives that the Nazis had developed up to 30 prototypes of antigravity craft, some of which were capable of space flight. These operatives also learned about the treaty between the Nazis and the Draco reptilians.

It was Admiral Botta's duty to understand both the war and post-war potential of the Nazi aerospace programs, and disseminate the relevant technical information acquired by the spies to his Navy superiors, and also to select U.S. aerospace companies, think tanks and government laboratories.

Tompkins job was to design intelligence briefing packets based on the Navy spy's debriefings, and to then deliver these packages to these leading U.S. aerospace corporations, think tanks and universities for study and evaluation.

Tompkins describes how from 1942 to 1945, he participated in over a thousand debriefings of U.S. Navy spies, who had been embedded in leading Nazi Germany aerospace corporations involved in building flying saucers.

The spies said that the Nazis had established two separate antigravity research programs. The first, located in Nazi Occupied Europe, was run by the Nazi SS and aimed at weaponizing antigravity prototypes in order to win the war. These were part of Hitler's promised wonder weapons that he believed would ensure Nazi Germany's eventual victory.

The second program, located in secure facilities under the Antarctica ice shelf, was primarily aimed at building spacecraft capable of interstellar travel. This more secretive program was led by German Secret Societies whose antigravity research had begun in the 1920's.

Disseminating this vital intelligence information to various U.S. aerospace corporations and universities was Tompkins primary responsibility.

Tompkins was also sent to flight school and later flew most of the Navy aircraft as a pilot, often flying admirals out to places like Douglas Aircraft in Santa Monica, Long beach and China Lake.

Between 1942 and 1947, Tompkins had access to highly classified projects and was involved in some of the most unprecedented advanced scientific programs on the planet. At this time, the infamous Philadelphia Experiment also occurred. William Tompkins and Admiral Rico Botta, were getting intelligence about this experiment gone wrong. And Tomkins stated that Admiral Botta attended several secret meeting, without Tompkins himself present, concerning this experiment.

After the official end of WWII, Admiral Rico Botta went on to play a key role in starting a U.S. Navy led secret space program filling a number of positions until his final "official" assignment at Naval Air Material Center, Philadelphia, from 1950 until his retirement in 1952.

Meanwhile, Secretary of the Navy, James Forrestal, commissioned the highly above top secret, think tank, project RAND in December, 1945 to study the implications of threatening extraterrestrial agendas.

This was because of U.S. recovered crashed extraterrestrial UFOs in 1941, at Cape Girardeau, Missouri, as documented in the book *MO41 The Bombshell Before Roswell,* by Paul Blake Smith, and in 1942, another two recovered UFOs.

Tompkins revealed that in early 1942, after the Battle of Los Angeles UFO incident, the President of Douglas Aircraft Company, Donald Douglas, Snr, along with his chief engineer, Arthur Raymond, and his assistant, Franklin Bolbohm, convened an informal working group that included two Navy admirals and two Army Air Force Generals to investigate the research implications of two retrieved UFO craft.

A leaked Majestic Document dating from February 1942, supports Tompkins claim that two UFOs had been retrieved after the Los Angeles incident, by the Navy and the Army respectively:

> "Regarding the air raid over Los Angeles, it was learned by Army G2 that Rear Admiral Anderson has informed the War Department of a naval recovery of an unidentified airplane off the coast of California with no bearing on conventional explanation.
>
> Further, it has been revealed that the Army Air Corps has also recovered a similar craft in the San Bernardino Mountains east of Los Angeles which cannot be identified as conventional aircraft. This Headquarters has come to the determination that the mystery airplanes are in fact not earthly and according to secret intelligence sources they are in all probability of interplanetary origin."

Also the Naval Intelligence data coming in from Germany on Nazi secret flying saucer projects and extraterrestrial treaties contributed to the decision to create the RAND project.

After World War II, in 1945, when U.S. authorities thought that the militarily defeated Germans might be willing to cooperate, they organized a secret mission to the German base in Antarctica.

The pilot chosen for this secret mission was Naval Commander, Graham Bethune. Bethune had a top secret security clearance, had graduated from Academy Air at Pensacola, Florida in 1943, and was an excellent celestial navigator, He also had a lot of night flying experience, hunting German Submarines during the war.

Naval Commander, Graham Bethune, flew Admiral Richard E. Byrd, along with British and American scientists to Antarctica to meet with German scientists as an attempt to have the Germans reveal the secrets of their flying saucers. This mission was a total failure as the Germans refused to reveal their secret technology. Commander Bethune then flew Admiral Byrd back to the Pentagon to report this matter. As a result of the lack of cooperation by the Germans, Operation High Jump was planned.

Operation High Jump was disguised as a scientific expedition, but was actually intended to destroy the secret German Nue Berlin base. This 1947 operation was a total failure after coming under attack by German flying discs and laser ray guns, which caused heavy losses to the men, aircraft and ships of Operation High Jump. Much of Operation High Jump is still highly classified today.

After the failure of Operation High Jump, President Truman was considering destroying the Nazi Antarctic base with a nuclear bomb. The Paperclip Nazis brought into the U.S. after the War and hired by U.S. corporations still retained their loyalty to the, now secret Nazi cause or the Nazi International that researcher, Joseph Ferrell has uncovered.

These Paperclip Nazis got wind of Truman's plan to nuke the Antarctic base and informed their compatriots down there. This led to several massive overflights of Nazi UFOs, launched from Antarctica, over Washington D.C. in 1952. Fighter jets were scrambled to intercept these UFOs but they were totally out maneuvered and helpless to defend the Capitol.

This show of force by the Nazi International let Truman know that the U.S. Capitol could also be destroyed with their superior technology. This led to secret negotiations between the Truman Administration and the secret Nazi International. Truman did not like or agree with all the Nazi International demands, but did agree to leave their Antarctic base in peace.

The negotiations continued into the Eisenhower Administration. One interesting meeting between the Nazi International and Eisenhower took place at Holloman AFB, near White Sands Proving Ground in New Mexico, on February 11, 1955, according to NASA employee, Clark McClellan, who got the information from German Scientist, Ernst Steinhoff.

Steinhoff had worked at Peenemunde in charge of the V-2 Program and submarine launched missiles. After the war, he was brought to the U.S. in operation Paperclip and in 1949, started working at the White Sands Proving Ground and Holloman Air Force Base.

Air Force One, a Lockheed Constellation, named *Columbine III* and carrying the U.S. President, landed at Holloman at 9:00 AM on February 11, 1955. The plane stopped in the middle of the runway and parked there. All Radars were ordered to be turned off. Soon, two flying saucers were visually spotted approaching the airbase.

One saucer landed opposite Air Force One, while the other remained in the air, observing events below. Eisenhower disembarked from the *Columbine III* and walked over to the landed saucer and entered it, according to witnesses of the event. 45 minutes later, he departed from the saucer and returned to his plane which then proceed to the air base terminal. The President stayed at Holloman until about 4:30 or 5:00 PM visiting with

Air Force personnel, and then left. (3)

According to what McClellan was told by Dr. Ernst Steinhoff, who was present at Holloman at that time, the saucers were the German V-7s, which the Nazi had improved since they were first operational in 1942 at Peenemunde and it was German officers – not extraterrestrials, as some have supposed – that Eisenhower was meeting with!

In a later meeting on February 1960, President Eisenhower had traveled to San Carlos de San Bariloche, Argentina, where officially he negotiated the Joint Declaration of Bariloche with Argentinian President, Frondizi concerning Peace and Freedom in the Americas.

However, San Carlos de San Bariloche, at the time, was the secret headquarters of the Nazi International and Adolf Hitler and his wife, Eva lived nearby at the Inalco estates. The real topic of negotiations concerned deals which would further put the U.S. Military Industrial Complex firmly under the control of the Nazi International.

Finally, after much negotiation, a secret treaty was agreed to. This treaty basically handed over to the Nazis, the U.S. Military Industrial Complex. This basic sellout to the Nazis by Eisenhower was supposed to be a stalling action until the U.S. could develop equivalent technology and have a fighting chance against the Nazis.

But, the Nazis were technically far ahead of the U.S. And, these Nazis were also taking over the CIA and NASA. By 1960, the silent and secret victory of the Nazis over the U.S. was complete. Eisenhower tried to warn the American people about the Military Industrial Complex in his farewell speech. But, the people just didn't understand the danger from this hidden direction. And, government secrecy did not help matters either!

President John F. Kennedy became the next President and became aware of the problem. A week before the JFK assassination in Dallas, JFK had met with NASA employee, Clark McClelland at Cape Canaveral and told him:

> "There is a plot in this country to enslave every man, woman and child. Before I leave this high and noble office, I intend to expose this plot."

But, the Nazi International, using the CIA, assassinated Kennedy before he could do anything meaningful.

Project RAND was set up as a highly secret contract to Douglas Aircraft in a highly classified area of Douglas engineering section at Santa Monica. Also at the same time, a new office of the Deputy Chief of Air Staff for Research and Development – to which RAND reported - was set up with Major General Curtis LeMay as head. On March 2, 1946, RAND was placed under the direction of Douglas's Assistant Chief Engineer, Frank Colburn - giving birth to the Douglas Think Tank, the Advanced Design Section. Frank Colburn had already been researching all he could about UFOs since the battle of Los Angeles, which he had personally witnessed in 1942.

RAND had two missions: (a) to research the potential design, performance, and uses for manmade satellites; (b) to function as a highly classified scientific research program. The latter mission also considered how to effectively deal with extraterrestrial technology that was thousands of years ahead of ours.

On September 24, 1947, MAJESTIC-12 was created upon the

recommendations to President Truman, from Sectary of Defense, James Forrestal, Director of the Office of Scientific Research and Development, Dr. Vannevar Bush, and Admiral, Roscoe H. Hillenkoetter. These recommendations were prompted by the Roswell and Corona, NM crashed UFO recoveries, as well as the earlier UFO recoveries, which were being examined at the Alien Technology division at Wright Patterson Field in Dayton Ohio.

MAJESTIC-12 was a top secret extraterrestrial related research and development program. The Deputy Chief of Air Staff for Research and Development and RAND both reported to MAJESTIC-12 after its creation. Later, in the Eisenhower Administration, on the advice of the Rockefeller commission to preserve secrecy, MAJESTIC-12 was placed under the control of the CIA. After that, even U.S. Presidents had little knowledge of what MAJESTIC-12 was up to!

On November 1, 1948, the Project RAND contract was formally transferred from the Douglas Aircraft Company to the RAND Corporation, due to conflicts of interests. Project RAND separated from Douglas Aircraft and became the nonprofit RAND Corporation. Some Scientists and Engineers from Douglas went to Rand Corporation and others stayed at Douglas, working at the highly classified, Advanced Design Section of Douglas.

The Navy one stayed at Douglas - even though Douglas was in the aircraft business. And the Air Force one moved to the new RAND location.

Former Secretary of the Navy, James Forrestal, who became the inaugural Secretary of Defense in September 1947, was locked in a number of bitter policy struggles with Stuart Symington, the first Secretary of the U.S. Air Force.

Symington was firmly opposed to Forrestal's policies, and was a direct factor in the events that led to his sacking as Secretary of Defense on March 28, 1949. One issue was that RAND, which Forestall originally created as Secretary of the Navy, was now favoring the Air Force. It was basically a turf war between the Air Force and Navy and their separate secret antigravity space programs.

And significantly, this was the same period when the Navy and Army Air Force collaboration in Project RAND came to an end, and the RAND Corporation was launched under Air Force dominance in early 1948.

These events led to the evolution of two separate and secret space programs – one run by the Navy and one run by the Air Force.

In 1949, Tompkins started working at Douglas Aircraft corporation. Because of his high security clearances with the Navy, he was placed into Advanced Design Section of Douglas. According to Tomkins, this secret Think Tank conceived and planned not only the Apollo Manned Moon Mission 4 years before NASA even existed, but also manned planetary missions within our solar system and manned missions to 12 of our closest stars!

As Engineering Section Chief, he conceived dozens of missions and spaceships designed for exploratory missions to the planets that orbit our nearest stars, a station to be built on Mars, a 2,000-man military base on the Moon, and 600-man Naval station for all the habitable planets and their moons.

Dr. Robert M. Wood helped William Tompkins edit his book, "Selected by Extraterrestrials", and also worked at Douglas

Aircraft at the same time as Tompkins, and although he didn't know Tompkins at the time, he knew some of the same people that Tompkins worked for.

With a Bachelor of Science in Aeronautical Engineering from Colorado and a Ph.D. in physics from Cornell, a 43-year career at McDonnell Douglas managing research and development projects and over thirty years of investigation into UFOs, Dr. Robert M. Wood is uniquely qualified to provide credible analysis about the nature of the UFO reality.

Obviously, obsolete rocket technology was not to be used on these missions to the stars. Scientists working at Advanced Design knew about the missing terms in Maxwell's equations on electromagnetic interactions. When these missing terms were corrected in the 1950s, the doorway to the stars were opened.

Also, according to Tompkins, Edward Teller also discovered a problem with a time factor in Einstein's $E = MC^2$ or $E = M (9 \times 10^{16} \text{ Meters}^2/\text{Seconds}^2)$. This, Teller corrected in the 1950s. The corrections in Maxwell's EM theory and $E=MC^2$ provided the technology to far exceed the speed of light.

These statements correlate with statements of past head of Lockheed Martin's Skunk Works, Ben Rich. At an alumni engineering meeting at UCLA, in March 1993, Ben Rich said, while showing a slide of a black disc heading towards space;

> "We now have the technology to take ET home. No, it won't take someone's lifetime to do it. There is an error in the equations. We know what it is. We now have the capability to travel to the stars. First, you have to understand that we will not get to the stars using chemical

propulsion. Second, we have to find out where Einstein went wrong."

According to Bill Tompkins, a Navy space craft could leave the Earth and be within the center of our galaxy in about 45 minutes! At Advanced Designs, Bill Tompkins was designing Naval space craft that were about 2 kilometers long for the Naval space program. This program was designed for the Navy – not the Airforce.

This was because the Navy had centuries of experience with voyages that would take months to complete. The planning, provisioning for manned crews living in close quarters for months at a time and so forth was a Navy thing – not an Air Force one.

As it turns out, contrary to Einstein, there are things that travel much faster than the speed of light. Einstein needs to be updated at our universities.

One of these things are scalar electromagnetic waves. As revealed by Thomas Bearden, scalar electromagnetic waves exist only in the time domain and have no extension in space. In the time domain, each instant is present everywhere in the universe simultaneously. Therefore, scalar waves can instantaneously be transmitted to any location in space.

Other things are torsion waves, generated by spinning objects. Russian scientist, Dr. Kozyrev, experimentally determined that torsion waves travel at a speed 9 orders of magnitude faster than light - if not instantaneously.

Also, apparently there are ways for matter to also travel at much greater than light velocities. One, of these methods use traversable wormholes for entry and exit points, some of which are naturally occurring.

According to Bill Tompkins, a natural traversable wormhole was first discovered in Northern Virginia and later, all over the planet. These are in use to this day. Also, not only artificial gravity, but also traversable wormholes can be created electromagnetically. The math theory of this is described in my book, *The Covert Colonization of Our Solar System* in the chapter on scalar physics.

Perhaps the most interesting thing William Tompkins discovered while working at Advanced Design in Douglas Aircraft, was that his secretary, a knockout gorgeous woman, wasn't human at all - but of the Nordic extraterrestrial race! She would never admit it. But, in additional to her female charms, she displayed great intelligence, leadership, and telepathic ability.

Quite often Bill, as she would call him, would hear her thoughts, planted into his mind. And, many of these thoughts provided the perfect resolution to the many problems in design and logistics for the Apollo, Moon and other Programs.

Later on, Tomkins would realize that the Nordics were helping the Navy space program and the Draco Reptilians were helping the Air Force space program. In effect there was a proxy war between the Air Force and Navy in these secret projects. Also, there seemed to be sabotage attempts on the Navy program which were coming from the Dracos, as outlined here: http://exopolitics.org/rand-corporation-part-of-alien-proxy-war-involving-usaf-navy-secret-space-programs/

Along with designing ships for this this secret space program, Tomkins was asked to design Deep Underground Facilities or

Deep Underground Military Bases (DUMB), Deep Under Sea Facilities and Submarine Launched Ballistic Missiles (SLBM) technology in the late 1940s and early 1950s.

Many in the Department of the Navy thought that SLBM couldn't be done. But, the Germans had already been designing and building SLBM subs at the end of the war.

Top military leaders wanted technology designed for the military so that they could match the extra and intra terrestrials (which reside inside the Earth) on an even playing field.

Also, Bill Tompkins was asked to design an underground tube tunnel system from the Pentagon to Edwards AFB in California and then on to Vandenberg. This system would propel high speed shuttles at supersonic velocities in a vacuum.

The Douglas Advanced Design insiders already were well aware of the extraterrestrial presence around the Earth. Not only were extraterrestrials flying their scout ships in our atmosphere from motherships further out in space, but intraterrestrials were living inside our planet in large underground caverns and in undersea bases which contained whole cities and civilizations unknown to surface humanity. Top people in Military Intelligence were "in the loop" on these extra and intra terrestrials, but it was decided early on that this information must be kept from the public because it would create a destabilizing influence on society.

For these reasons, the Douglas Advanced Design Think Tank people were very careful to conceal the ET subject from the regular Douglas Corporate managers. Bobby Ray Inman was cleared for ET knowledge and acted as a "go between" Advanced Design and Secretary of Defense, James Vincent Forrestal.

Although Forrestal had been opposed to the National Security Act of 1947, and also opposed any form of unification of the Navy with the Army, he had accepted the job of Secretary of Defense at the newly built Pentagon in September of 1947. Secretly, he was also made an original member of MJ-12 by President Harry Truman around the same time. However, he personally disagreed with all the E.T. secrecy and felt that the public had a right to know what was really going on.

His secret membership in MJ-12, which oversaw the ET project and all the corresponding secrecy surrounding that project conflicted with his personal feelings that the public had a right to know about the ET presence on the Earth. And, this began to cause a lot of stress and inner turmoil in his life.

Also, politics entered the scene, As the presidential election grew closer, Forrestal held some confidential meetings with Governor Thomas Dewey, Truman's opponent. These meetings were designed to ensure that Forrestal would continue at the Defense Department if Dewey defeated Truman, as many expected.

Columnist, Drew Pearson exposed the meetings some time before the election. The exposure strained already difficult relations between Truman and Forrestal, and in March 1949, Truman abruptly requested Forrestal's resignation as Secretary of Defense. His membership in MJ-12 was probably also terminated at the same time.

On the very day of Forrestal's removal from office, he allegedly went into a deep depressive state. He was flown to Florida where his wife was vacationing.

Dr. William Menninger diagnosed Forrestal as severely

depressed, but rather than having Forrestal admitted to his own clinic, one that was familiar with such conditions, Forrestal was flown back to Bethesda, Maryland and committed to the VIP suite of the National Naval Medical Center.

This was an odd destination, considering that Forrestal was no longer a government employee. Even stranger was the fact that access to him was severely limited, something the government denied, but that Forrestal's brother, Henry, confirmed.

The Truman Government was concerned that Forrestal would go public with the MJ-12 operation and the government's secret dealings with E.T.s.

Basically Forestall was murdered on May 22, 1949 and it was made to look like a suicide by jumping out a window on the 16th floor of the VIP suite at the Hospital.

That gives you a good idea of what lengths the government would go to, in maintaining secrecy about E.T.s and the secret space program.

Nevertheless, secretly, the Nordics extraterrestrials were helping the U.S. secret government take the first steps into space. And, William Tompkins was right in the middle of the planning department for this secret space program! (4)

Other Science

From the Nineteenth century to the end of World War II, scientists in the area now known as Germany, Italy and Austria were developing what could only be called "Other Science". As we all well know, alchemy, which led to many important chemical discoveries was finally discredited by most contemporary scientists because it was clothed in mystical language to hide its secrets. However, alchemy was a secret art and number of well documented cases, witnessed by credible people of the actual alchemical creation of gold have occurred in the past.

Baron Dr. Carl (Karl) Ludwig von Reichenbach discovered paraffin in 1830, one practical result of his own research with coal tar and coal tar derivatives. He did not stop making chemical discoveries of commercial impact however. From coal tar he extracted the antiseptic Eupion (1831), the preservative and therapeutic agent Creosote (1832), the indigo dye Pittical (1833) and Cidreret (a red dyestuff), Picamar (a perfume base), as well as Kapnomor, and Assamar. The successful commercial development of these organic substances brought him into greater wealth. Reichenbach's discoveries founded the huge dye and chemical industries by which Germany made legendary fortunes, which few but German chemists remember.

Very gradually distinguishing himself as an exemplary industrial engineer, he began establishing ironworks (Villengen, Baden), charcoal furnaces (Hausach, Baden), metallurgical and chemical works (Blansko, Moravia), steelworks (Turnitz, Austria), and blast-furnaces (Gaya, Moravia). His wealth increasing beyond all reckoning, he purchased lands literally from the Danube to the Rhine. His fame and reputation as an industrialist and research scientist spread across Europe. In short, he was an exemplary scientist-mogul of legendary proportion.

But, Reichenbach's most interesting discoveries are not known about or taught in the Universities today.

He discovered that sleepwalkers were dramatically affected by moonlight - particularly during the full moon. If moon light was shielded from their homes, there were much less incidents of sleep walking among the patients.

He employed psychic sensitives in his research, as there were no existing instruments to measure the energies that he was studying. He would expose a plate of metal to the light of the full moon and then bring a conducting bundle of braided wire electrically connected to the plate into a room shielded from moonlight.

When this wire was brought close to these sensitives in the room, they would experience muscle spasms and nausea. Less sensitive persons would have to actually touch the wire to get the same effect. There were no deflections of compass brought near the wire during these experiments and when touched to an electroscope there was no deflection of the leaves. So, the energy was not electrical.

Similar experiments, using sunlight in place of moonlight, would cause different results. The sensitives sitting in darkened rooms, felt invigorated when the wire was brought close to them.

They could tell the difference in the effect when prisms were used to divide sunlight into its spectrum. The violet light was invigorating and uplifting while the red part of the spectrum also seemed invigorating, but in an irritating way to these sensitives.

He could place metal plates out in sunlight connected by a wire to similar plates in a darkened basement suspended over a pan of dirt with plant seeds. In this manner, the plants would grow even in complete darkness!

In another experiment he was able to determine the speed at which this energy traveled along the wire. It was about the speed of a man walking. He called this newly discovered life force energy, Od. Metal was a good conductor of Od while organic fabrics were insulators of Od.

In experiments with magnets and sensitives in darkened rooms, the sensitives could see the emanations from the magnetic poles. These emanations looked like blue flames from the north pole and like red flames from the south pole. When copper wire was wound around the magnet and the end of the wire was led into another darkened room, sensitives could see flame like emanations from the wires end.

A good description of Reichenbach's work is contained within Gerry Vassilatos' book, *Lost Science*.

So here, we see an example of "Other Science", not taught in our universities. Another example was Wilhelm Reich, who originally lived in Germany but migrated to the U.S. when Hitler came into power. He had discovered a similar energy to Od, which he called Orgone. This energy could be used for many things including to control the weather and cure cancer. At one point in his experiments using Orgone, Wilhelm Reich created living biomes from totally inorganic material like sterilized silica sand.

For his efforts in the U.S., he was imprisoned and his books burned by a court order. Shortly before he was to be released from prison, he mysteriously died. The Medical establishment was making too much money in cancer research and definitely did not want to see an actual cure. It seems that Wilhelm Reich just couldn't escape the forces of fascism.

But, his students have carried on his work which is described in many books and on the internet. One who studied Reich's work was James Trevor Constable. He developed Reich's Cloud Buster into improved designs and started his own weather engineering company using the concepts of both Orgone and ether: http://www.rainengineering.com/

In Austria, Victor Shauberger was a forestry warden. He closely observed nature and developed his own water science which led to improved logging flumes and water treatment devices. This science led to developing his *Repulsine* and *Trout Turbine,* based on his close observations of trout swimming up waterfalls. These machines had both free energy aspects and levitation aspects.

Victor Shauberger privately met Adolf Hitler in 1934 to discuss the fundamental principles of agriculture, forestry, and water engineering. While Hitler was impressed by Schauberger's radical ideas for utilizing water power in new and dynamic ways, he was also displeased that Schauberger was not willing to participate in work for the Third Reich.

This all changed when the Nazis annexed Austria in 1938 and war broke out in 1939. The SS would come searching for him. Soon, he was forced to work for the Nazis. Victor Shauberger's trout turbines were first considered by the Nazi SS for submarine propulsion and then later for flying saucer propulsion.

At Mauthausen, under orders from Heinrich Himmler himself, Schauberger was to carry out research and development for the Third Reich war effort. He was given approximately 20-30 prisoner engineers to proceed with his research into what was termed "higher atomic energies". For this Schauberger was given special dispensations from the SS for both himself and fellow engineers.

The construction and perfection of the Repulsin A model discoid motor continued until one of the early test models was ready for a laboratory test that ended in disaster. The model was 2.4 meters in diameter with a small high-speed electric motor. Upon initial start-up the Repulsin A was set in motion violently, sheared the anchor bolt holding it down, and rose vertically, quickly hitting the ceiling of the laboratory, shattering to pieces. The SS were not pleased and even threatened Schauberger's life, suspecting deliberate sabotage. But, the workers had merely failed to recognize the tremendous lifting forces that would be generated by the device and had tested it in Shauberger's absence.

Replacement models were built, but by 1943 a more improved design, the Repulsin B model was constructed with the SS objective of developing this motor for an odd SS bio-submarine which Schauberger named the "Forelle" (Trout) due to its configuration of a fish with a gaping mouth.

Work on the Repulsin B continued in 1944 at the Technical

College of Engineering at Rosenhügel in Vienna. Schauberger was finally released back to Leonstein, Austria that same year. It appears that the SS had discarded the idea of applying the Schauberger motor to a submarine when the benefits would greatly improve their work on the secret Flugkreisel which was taken from Rudolf Schriever back in 1941. By 1943 the machine had flown but proved to be unstable.

The leader of the SS replacement team was Dr. Richard Miethe who proposed several Flugkreisel replacements with varied power plants, most of which relied on jets or rocket power, until it was learned that Schauberger had engineered a type of turbine machine that would create an up-current of axially-spinning air so powerful that the up-current's drag force would speed the whole machine higher and higher into the air with a thrust equal to 10,000 hp simply by moving "air".

This is how Schauberger described the working principals:

> "The principal of the vacuum turbine consists in the fact that that an artificial thunderstorm or high tension electrical charge develops in a capillary tube and a double spiral pipe in which electromagnets have been incorporated, which transforms the substance of the air into electrical energies.
>
> These discharge into the sidewalls and from there thru diffusion are ducted away downwards as heat. Thru the transmutation of aeriform matter into energies, a self-intensifying physical vacuum is created in an upward direction and through the recurrent development of a heat gradient in a downward direction; a powerful up current of air evolves which after a few minutes intensifies to cyclonic force. Because the air pressure in an

upward direction can be progressively reduced in this extremely simple way, the counter flow of air can be raised to over 124 mph producing about 10,000 brake H.P. on the inclined surfaces of the prop or turbine. If water or air is rotated into a twisting form of oscillation known as 'colloidal', a buildup of energy results, which, with immense power, can cause levitation. This form of movement is able to carry with it its own means of power generation. This principle leads logically to its application in the design of the ideal airplane or submarine... requiring almost no motive power."

The turbine was considered a priority for flight development into a manned machine by the SS. It is speculated that Miethe's final design built in Breslau that flew in 1944 was an enlarged manned Repulsin-type craft.

The *Repulsine* levitator, is described in the book *The Energy Evolution* by Callum Coats and here: http://discaircraft.greyfalcon.us/Viktor%20Schauberger.htm

A most interesting development, however was the physics of the primary state of matter developed by Dr. Karl Schappeller. To get an idea about the primary state of matter, you would have to accept the presence of ether pervading all of the universe.

This ether not only is the conveyor of action at a distance via things we call magnetic, electric and gravitational fields but also is endowed with consciousness. This is the primary point of departure from conventional or secondary physics. This secondary physics represent the effects of the primary physics which causes the secondary physics.

This everywhere conscious ether is a property of the God of all that is. As this God is everywhere, this consciousness is everywhere in the ether. This ether imparts consciousness to all forms of matter which is formed from this ether, to greater or lesser degree.

Erwin Schrodinger, developer of the Schrodinger equation of quantum physics, stated it this way:

"The total number of minds in the universe is one"

This implied that the Consciousness of God is the source for all consciousness. Because of this, all is connected in ways that we can hardly imagine! And this is the reason behind psychic phenomena like telepathy and clairvoyance.

Also quantum mechanics implies that a particle must be consciously observed before it becomes localized. This correlates with statements by Seth [from the book *Seth Speaks* by Jane Roberts] that our material world is created by our consciousness.

One question we might ask ourselves is, "Where is the dividing line between conscious and unconscious matter?" Modern science teaches that our consciousness is created by our sensory input into our brain. However, the single celled amoeba knows in which direction to move to find food. It does this without a nervous system, eyes or a brain. How is the amoeba conscious of its surroundings?

We already know that plants are conscious, from many experiments done with them. What about minerals? Well, some healers know that crystals hold memory and can be programmed and "cleared". The same can be said for water.

When considering the atomic level, according to the Pauli

exclusion principle, no two electrons in an atom can be in the same quantum energy state. So how do these electrons know what all their neighboring electrons are doing so that they can do something different? These electrons seem to be aware of their surroundings, which is consciousness. So logically, it is not too difficult to see that all matter is conscious to some degree.

Karl Schappeller theorized that the sun was made of glowing magnetism. This glowing magnetism was about one thousand times stronger than normal magnetism. He even developed a spherical device which created this glowing magnetism in the center and created free energy and levitation.

A book, written in terms used in the early 1900s, that explains part of this "other science', the primary state of matter and the glowing magnetism of Karl Schappeller is Cyril W. Davson's *The Physics of the Primary State of Matter*

Also, the internet has a number of sites that try to explain the Schappeller device. Here is a good one: http://rexresearch.com/schapp/schapp.htm

So, we see that a science was being developed in this area of the world that was considerably different than the science taught in the universities then or today. This science was the "other science".

There is a large new development in "other science" in more modern times. This "other science" is being developed by nuclear physicist, Mehran Tavakoli Keshe.

As a young man, Keshe attended Queen Bay College in England, where he graduated as a nuclear engineer, specializing in reactor technology.

During this time, he began playing with his own theory of the creation of gravity. All he had at that time was a theory saying gravitational fields are created by the interaction of matter, dark matter and antimatter.

He knew his theory was correct because he could *see it working* in his mind's eye. In the same way, Nikola Tesla was able to *see* and understand the workings of the AC electric generator (the system that powers the world) in his mind's eye before ever constructing it.

As a young, ambitious scientist fresh out of school, there was not much opportunity for Keshe to gain employment doing research into his outlandish theory of the creation of gravity. So, he began his research as an independent scientist, free to do his own investigation, and free from the distractions of the mainstream scientific community.

It was a very long and difficult process, but Keshe was able to independently finance his research and develop his theory to the point where he could prove it scientifically by replicating the exact gravitational and magnetic field conditions of the Earth in a small scale system.

In his book, *The Universal Order of Creation of Matters*, Keshe explains the basic principles of how matter is created and what causes gravity. These are also the same principles that apply to the interactions within a neutron – interactions between Matter, Antimatter and Dark Matter. Keshe had to invent new words and concepts to explain his theory. He coined words like mafs, pimtics, magravs, and plasmatic magnetism fields.

Keshe states that the magnetic field is the ultimate building block of matter. This is similar to the concepts developed in the chapter on scalar physics in my book *The Covert Colonization of Our Solar System* with the difference that the scalar super potential [from which everything else is derived] was formed not of magnetic fields [Webers/ Meter squared] but of just Webers.

Plasmatic magnet fields, which are different from magnetic fields because they are detached from their source are called "pmtics". These pmtics can arrange their selves into circulating or vortexial configurations in different ways that give rise to the three types of mafs; matter mafs, antimatter mafs and dark matter mafs.

According to Keshe, the neutron is composed, not of three quarks, but three fundamental matters, he calls matter mafs, antimatter mafs, and dark matter mafs. The neutron is the primary plasmatic matter and decays into the electron and proton, each of which also contains the above three matter mafs in lesser quantities.

Kesha also states that the magnetic and gravitational field are always interlocked into a Magravs field.

I won't go into the complete theory, which was difficult for me to understand, because it was much different than my traditional training in physics and electrical engineering. I believe that it would be easier for someone starting out without the formal scientific training to understand his theory. Read his books and papers to get a better idea of his revolutionary theory.

The important thing is that Keshe has created working technology based on his theory that can generate free energy, antigravity, nanotechnology, remove CO_2 from the atmosphere, neutralize radioactivity, reduce pain and so on. Keshe created the Keshe Foundation to educate the people of the planet into this new science and technology. www.KesheFoundation.org

The Keshe Generator is a self-nourishing system that consists of a rotating plasmatic magnetic field within a coil of copper wire, and has been developed from our new understanding of the creation of gravity. Just as our electrons spin naturally within our atoms, so too does the Keshe Electric Generator spin naturally within the laws of nature.

With one spherical plasma generator, which weighed about 10 Kilograms, while under test levitated off the ground. It was actually able lift about 100 Kilograms in further testing. This result was not unexpected and gave experimental proof of Keshe's theories as explained in his book *The Universal Order of Creation of Matters*.

In 2008, the M.T. Keshe, a native of Iran, educated the Iranian Government on his Space Technology and gave a full transfer of knowledge to Iranian scientists. In 2011, Iran became the first nation in the history of the world to announce they have an established Spaceship Program based on Keshe's science. Fourteen nations have received a full transfer of this knowledge: Iran, India, Italy, Japan, Armenia, Sudan, Thailand, Bangladesh, Sierra Leone, Australia, Congo, Russia, China, and the United States of America. More nations are being added to this list regularly.

Unlike these conventional space programs, however, a Spaceship Program is developed around the use of Magrav Technology, using magnetic and gravitational fields as a mean of lift, drive and energy generation, as opposed to the use of combustion technology which will soon become obsolete.

A European Patent on Keshe's Magrav antigravity and free energy generator was filed on 3/10/2005 and Patent # EP 1 770 717 A1 was published on 4/4/2007. This patent should be studied because it gives a detailed course on plasma magnetic fields and Magrav fields and how they can be used to create antigravity and free energy devices.

I was able to obtain this patent along with a lot of other informative material from the Keshe Foundation like the design concept for their space ship and a number of scientific papers published by Nassim Haramein and Elizabeth A. Rauscher on the source of spin in the universe, the real strong force is gravity and other interesting leading edge physics subjects by becoming a Founding Member of *The New Energy Industry* for about $20 here: http://www.thenewenergyindustry.com

Keshe technology was developed into military applications by Russia and Iran and used against U.S. military weapons in several well-known incidents.

On 4 December 2011, an American Lockheed Martin RQ-170 Sentinel unmanned aerial vehicle (UAV) was captured by Iranian forces near the city of Kashmar in northeastern Iran. The Iranian government announced that the UAV was brought down by its cyberwarfare division. Actually, the Iranians recovered the U.S. stealth drone flying over their territory using Magravs technology given them by Keshe. When the U.S. requested "their toy" be returned, the Iranians politely refused.

On April, 2014, the USS Donald Cook, while performing a mission in the Black Sea substantially violated Russian Federation territorial integrity, causing President Putin to authorize an SU-24 electronic warfare defense attack against it that was so devastating this top-line warship was left helpless. Once back in its Romanian port, its American sailors were so demoralized that 27 crew members immediately filed letters of resignation from the U.S. Navy.

One year later, in May 2015, again in the Black Sea, the Arleigh Burke-class guided-missile destroyer USS Ross DDG-71 again substantially violated Russian Federation territorial integrity, causing President Putin, once again, to authorize an electronic warfare defense attack against it, but when confronted with the fate that befell the USS Donald Cook the year prior, the USS Ross retreated on seeing incoming SU-24s.

In March 2015, Russian Federation submarine naval forces attacked with their advanced electronic weapons the American aircraft carrier USS Theodore Roosevelt causing it to flee to the coast of Britain where it remained for weeks to the amazement of the public until it was repaired and able to get under sail again.

Russia's new electronic warfare defense are based on what is called Magravs technology (Western code name code name "Khibiny") developed by Iranian-born nuclear engineer Mehran Keshe - a modern day Nikola Tesla.

In 2012, Keshe wrote a personal letter to President Obama and warned:

> "...The aircraft carriers of the U.S. will become nothing but floating bathtubs if our Magravs technology is used effectively, and the runways full of F-16s and 18s and so on will be nothing but runway museums of iron birds, as these craft will not be able to fly if their electronic systems are once touched by Magravs space technology. These crafts and battle ships would have to be rewired from A to Z before they could ever operate again..."

Actually, Keshe has offered his technology to the government and companies of the United States on multiple occasions and was ontinually blocked. Such is the nature of an "exceptional country".

The Keshe Foundation's new (as of mid-2016) online store has a 3 Kilowatt Magravs free energy generator, that has no moving parts and uses no fuel, for sale. The interesting thing about these generators is that they are powered by plasma – not electricity. The plasma seems to know what is required at the load and provides it. The result is that the same generator will work on the European 220 VAC devices or the U.S 110 VAC devices or even on DC devices. They are for sale here: https://store.keshefoundation.org/store/product/magravs_power_generator/

The Keshe Foundation announced in April of 2012 that they are establishing the Spaceship Institute. An independent non-profit educational organization that aims to give mankind the real freedom to travel in Space, to explore Space in an easy inexpensive way, while at the same time realizing world peace. The Spaceship Institute is an open organization, accessible to all scientists and students of all nations, and is a subsidiary of The Keshe Foundation. They will begin by accepting one hundred (100) scientists from around the world to be educated on this new knowledge, so they may take this knowledge back to their own nations.

The Keshe Foundation is a Space Organization. The reason why they have pushed ahead and begun developing the medical aspects of this technology is, like any other space agency, they have to cover all aspects of space travel – medical, energy, food and water, and even protection against harmful cosmic radiation. All of these things must be understood when traveling in deep space.

The entire concept of the medical aspect of this technology is to develop one holistic medical application that is able to process and reset any illness that we currently know of, as well as any new viruses or diseases we may encounter in the future. This means, having one medical application that is able to reset one's body to its original, harmonious, balanced state of health.

As journeys in space will go further and last longer, space travelers will need a portable health system to help them recover from any type of illness. Human cells are made of a combination of plasmatic magnetic fields which form the structure of the atoms of hydrogen, oxygen, carbon and nitrogen that are the building blocks of the amino acids in the human body.

The Keshe space heath technology works directly with this plasmatic anatomic structure of the human body and shows it how to get back to its original state of health. The procedure works quickly with no tablets or injections and has no side effects. Numerous trials for this technology have been carried out with volunteers, with good results, who had exhausted all other medical options.

As opposed to the Secret Space Programs, which has been kept secret from the world's people, the Keshe Foundation Space Ship Institute (KFSSI) will hold no secrets back from the world's people. The technology is open sourced and available to all who wish to make use of it. They already have started a program called the MOZHAN program, which is looking for volunteers to become trained as astronauts: https://mozhan.kfssi.org/

The central plasma reactor besides providing lift and propulsion to the space craft also can be used to produce oxygen water and food for the space travelers. In their knowledge seekers workshop #2, the description and plans for a 40-meter diameter 5 stories high saucer shaped space ship is presented which they plan to have built by Italian factories.

The Space ship will have 4 plasma reactors arranged in a tetrahedral configuration. The top reactor will provide the primary lift MAGRAV field while the three lower reactors 120 degrees apart will provide the steering or navigation fields.

Keshe also claims that these plasma reactors also create protective fields around the spacecraft to deflect meteors and deflect harmful radiation away from the craft.

In Knowledge Seekers Workshop # 123 live stream video on July 26, 2016, Keshe said that the plasma in the reactor for the space ships or smaller air cars is conscious and intelligent and by placing your own amino acids from a drop of your blood into the reactor material, your imprint goes into the reactor. That will allow you to control the lift and direction of travel over the craft with your thoughts, emotions and intent!

This demonstrates the strong connection between consciousness and matter. According to a multidimensional entity named Seth, the material world is actually created by our consciousness, as described in *Seth Speaks* by Jane Roberts. This also demonstrates why people need to learn the ways of peace before going into space – especially in these type of space craft that can react to your thoughts and emotions!

In Knowledge Seekers Workshop #124 on July 28, we learn that the plasma energy from a MAGRAV generator powering 110 VAC electrical devices will not give you an electrical shock - unlike normal 110 VAC electrical supplies which will give you a good shock.

Also rotating spheres filled with Gans will cause rotation in water contained in bottles of water near (but not connected to) the rotating sphere. Gans are gases in nano solid state. Other workshops show how to create Gans.

Also rotating a Gans filled sphere in the middle of a MAGRAV "stacker coil" (which other workshops show how to build) caused a very personal enlightened states of mind in one experimenter – like becoming one with the universe.

Keshe is also modern alchemist. He claims that soon he will show how to produce gold on your kitchen table using this new plasma science. In fact, some of Keshe's students are already producing Gold Gans, or nano state gold, in their experiments. This new plasma science will definitely usher in many revolutionary new technologies in the future.

Not all persons and parties are happy with Keshe's discoveries which very much threatens to overthrow the status quo in many areas. Also, many scientists, trained to think a certain way, find it difficult to understand his plasma science, which nevertheless is experimentally verified.

There have been several unsuccessful attempts on Keshe's life and attempts to destroy the foundation by parties that feel their status threatened by his discoveries. But apparently, the forces of good have other plans because these attempts have all failed.

After he was poisoned in one attempt on his life, he cured himself using his plasma technology. The foundation and the teachings of the new plasma science is now international and would be just about impossible to stop now, even if Keshe was gone.

This new plasma science works similarly in creating galaxies, solar systems, human bodies, and atoms. And there are similarities to the Primary state of matter as outlined by Karl Shapeller, in that Keshe also states that this plasma energy is both conscious and intelligent.

When modern science finally understands that matter itself is conscious, it will go a long way in explaining mysterious quantum phenomena and include consciousness as another physics phenomenon.

"As a scientist, as a physicist, I treat the body as a galaxy. When you understand how a Galaxy works, you will understand how your body works." ~ M.T. Keshe

NASA

The National Aeronautics and Space Administration (NASA) is an independent agency of the executive branch of the United States Federal Government responsible for the civilian space program as well as aeronautics and aerospace research.

President Dwight D. Eisenhower established NASA in 1958 with a supposedly civilian, rather than military orientation encouraging peaceful applications in space science. The National Aeronautics and Space Act was passed on July 29, 1958, disestablishing NASA's predecessor, the National Advisory Committee for Aeronautics (NACA). The new agency became operational on October 1, 1958.

People who read the fine print of the NASA charter will discover that it is indeed connected to the military and is directed to keep secret any discovery that might impinge on U.S. National Security. For example:

> "Section 305 (i)…The National Aeronautics and Space Administration shall be considered a Defense Agency of the United States for the purpose of Chapter 17, Title 35 of the United States Code…

Section 206 (d) No [NASA] information which has been classified for reasons of National Security shall be included in any report made under this section...

These classified subjects include anything of extraterrestrial nature, including extraterrestrial civilization's ancient ruins, existing extraterrestrial bases, and recovered extraterrestrial technology.

Another facet of NASA is that it was run by former Nazis. This was the offshoot of Eisenhower's secret treaty with the Nazis that also gave them the control over the U.S. Military Industrial Complex, as revealed by Corey Goode.

This idea is also reinforced by the testimony of William Tompkins, who on visiting the Redstone arsenal observed an American flag with a swastika superimposed on it. Tompkins also said that the former Nazis were actually running NASA. He stated that NASA was basically a continuation of the Third Reich's own space program.

The American flag with a swastika superimposed on it is quite symbolic of U.S. foreign policy following World War II. Most true historians will remark on how militarily aggressive U.S. foreign policy became following that war. The Nazi takeover of the Military Industrial Complex required a lot of war making. Now, the U.S. is the world's greatest threat to world peace, as it has systematically destroyed one nation after another from Yugoslavia to Libya.

Most of these German scientists used in the military rocket programs and other weapons programs were brought into the U.S. after the defeat of Germany while the U.S. was still at war with Japan. The military wanted any weapons that might help them to defeat Japan and they needed the Germans in the U.S. to assist in the development of these weapons. Also, there were many that were concerned by the Soviet Union – a then present ally that could become a powerful enemy.

The War Department's Joint Intelligence Objectives Agency (JIOA), headed by Director Bosquet Wev, wanted these German Scientists brought to the U.S. But, there was a problem with the U.S. State Department and Justice Department, both which prohibited granting U.S. visas to any former members of the Nazi party.

When Wev submitted the first batch of German Scientist's dossiers to the State Department representative on the JOIA board, Samuel Claus, Claus claimed everyone was an ardent Nazi and denied the visas.

Wev then wrote a memo stating "The best interests of the United States have been subjected to the efforts expended in in beating a dead Nazi horse." He also declared that sending these German scientists back to Germany, where America's enemies could exploit them presented "a far greater security threat to this country than any former Nazi affiliations that they may have had or even any Nazi sympathies they may still have."

The JOIA board came up with a simple solution to the problem. They would simply "sanitize" the dossiers on any other German Scientists they wished brought in by removing any references to Nazi affiliations of these scientists. Paper clips were used to hold the sanitized record on top of the actual record, which caused JOIA to name their plan *Operation Paperclip*.

Unknown to the public, many thousand former Nazis were brought into the U.S. in *Operation Paperclip*. But, the Russians got even more of these scientists after the war.

NASA was basically, a military operation disguised as a civilian one. It was largely a "smoke screen" to divert public attention away from the secret space programs, which were developing the means to start exploring the universe. The public program used rocket technology to get into space while the secret programs used antigravity technology. Also recovered extraterrestrial exotic technology used in these secret space programs, was kept secret from the public and have them think that rockets were the only way to get into space.

As many researchers have discovered, NASA is notorious for hiding all evidence of extraterrestrial artifacts or other clues to their existence on the Moon or other planets. Exopolitics, or the interaction between Earth governments and extraterrestrial governments, was kept secret because it was feared by the powers that be, that public knowledge of the extraterrestrials would destabilize these governments control over the people of the Earth.

A number of persons have come forward to tell us what they know about the E.T. presence on the Earth and near Earth space being kept from the world's public.

Dr. John Brandenburg, PhD is a plasma physicist. He did his graduate work in California at Lawrence Livermore National Lab in controlled plasmas for fusion power, and has worked in energy, defense, and space research for a number of years. He is currently a consultant at Morningstar applied Physics LLC, and a part-time instructor of Astronomy, Physics, and Mathematics at Madison College.

He is the principle inventor of the MET (Microwave Electro-

Thermal) plasma thruster using water propellant for space propulsion.

Brandenburg was involved in the Clementine Mission to the Moon, which was part of a joint space project between the Ballistic Missile Defense Organization (BMDO) and NASA. The mission discovered water at the Moon's poles in 1994. He was the deputy manager of that mission.

He said that the Clementine Mission was basically a photo reconnaissance mission to check out if someone was building bases on the moon that we didn't know about. And, were they expanding them?

He also went on to state that after the completion of the Clementine Mission, the photographs were analyzed by an elite defense department team with the highest security clearance. Brandenburg had this to say about the photographs:

> "Of all the pictures I've seen from the moon that show possible structures, the most impressive is a picture of a miles wide recto-linear structure. This looked unmistakably artificial, and it shouldn't be there. As somebody in the space defense community, I look on any such structure on the moon with great concern because it isn't ours, there's no way we could have built such a thing. It means someone else is up there."

He's also been quoted as saying:

> "We were aware there was a possibility of an unknown presence, possibly alien/extraterrestrial near the Earth ... There I am sitting in a room of retired army and Air Force generals and a few admirals, and we're watching what

looks like a firefight in space. The most senior general there ... turned to me and said, 'Where do you think they're from?' and I said, 'I don't know sir, I've heard they're from 40 light years from here.'"

Here is what some other prominent persons have to say on the subject:

"Intelligent beings from other star systems have been and are visiting our planet Earth. They are variously referred to as Visitors, Others, Star People, ETs, etc. ... They are visiting Earth now; this is not a matter of conjecture or wistful thinking." — Theodore C. Loder III, PhD, Professor Emeritus of Earth Sciences, University of New Hampshire

"There is another way, whether it's wormholes or warping space, there's got to be a way to generate energy so that you can pull it out of the vacuum, and the fact that they're here shows us that they found a way." — Jack Kasher, Ph.D., Professor Emeritus of physics, University of Nebraska.

"There is abundant evidence that we are being contacted, that civilizations have been visiting us for a very long time. That their appearance is bizarre from any kind of traditional materialistic western point of view. That these visitors use the technologies of consciousness, they use toroids, they use co-rotating magnetic disks for their propulsion systems, that seems to be a common denominator of the UFO phenomenon." — Dr. Brian O'Leary, Former NASA astronaut and Princeton physics professor.

Some NASA employees were instructed to air brush out any alien ruins or existing bases from photographs taken from Lunar and Mars orbiters.

For example, Donna Tieze, a former photo technician at NASA's Johnson Space Center in Houston revealed that a coworker in a restricted area had the job of airbrushing out UFOs and anomalies, suggesting ancient ruins or existing bases, from NASA photos before they were released to the public. She made these revelations on May 6, 1995 in a radio interview on WOL-AM in Washington D.C. and later, at Steven Greer's Disclosure Project Press Conference in 2001.

NASA Astronauts have seen UFOs following them while in space but have been sworn to secrecy on these issues. And many have taken these secrets to the grave without talking.

However, there have been a few exceptions. Clark C. McClellan is one Space Craft Operator who has decided that the taxpaying citizens of the U.S., who actually pay for NASA, have a right to know what is really going on. He has an informative website here: http://www.stargate-chronicles.com/site/

For turning whistleblower, McClellan was stripped of all his benefits and pension from working at NASA and now is barely living off of social security.

For years the government has denied that an extraterrestrial space craft crashed at Roswell in 1947 and claimed it was just a weather balloon. But, McClellan just happened to be a friend of Dr. Wernher von Braun and at one time asked him about the Roswell incident.

At the time of the crash, Dr. von Braun, who would later become a director of NASA, was working at the nearby White Sands, New Mexico rocket test site.

Dr. von Braun explained how he and his associates had been taken to the crash site after most of the military were pulled back. They did a quick analysis of what they found. He told me the craft did not appear to be made of metal as we know metal on earth. He said it seemed to be created from something biological, like skin. McClellan was lost as to what he indicated, other than thinking perhaps the craft was "alive."

The recovered bodies were temporarily being kept in a nearby medical tent. They were small, very frail and had large heads. Their eyes were large. Their skin was grayish and reptilian in texture. He said it looked similar to the skin texture of rattle snakes he'd seen several times at White Sands.

His inspection of the debris had even him puzzled: very thin, aluminum colored, like silvery chewing gum wrappers. Very light and extremely strong. The interior of the craft was nearly bare of equipment, as if the creatures and craft were part of a single unit.

Dr. von Braun revealed this information to McClellan because he knew that McClellan had taken the secrecy oath and thought that he would never reveal it.

McClellan has other startling revelations. He personally observed an 8 to 9-foot-tall ET on his 27-inch video monitors while on duty in the Kennedy Space Center, Launch Control Center (LCC). The ET was standing upright in the Space Shuttle Payload Bay having a discussion with two tethered US NASA Astronauts. This Space Shuttle was on a secret military mission.

He also observed on the monitors, the spacecraft of the ET as it was in a stabilized, safe orbit to the rear of the Space Shuttle's main engine pods. He observed this incident for about one minute and seven seconds - plenty of time to memorize all that he was observing.

Other revelations of McClellan include that Dr. Kurt H. Debus worked on Die Glocke or Nazi Bell as well as Hans Kammler and that McClellan recognized Hans Kammler at NASA in Dr. Debus' office.

Die Glocke was an experimental space-time warping machine which the SS were testing as an antigravity propulsion machine to place in their Saucers. General Patton recovered one version of Die Glocke and sent both the Die Glocke and Kammler back to the Wright Patterson AFB. Kammler worked there for a number of years, then worked at NASA.

Kammler was ordered by Hitler to execute Peenemunde scientists to stop them falling into Anglo-American hands. It was a huge risk to Kammler's own personal safety to evacuate all the Peenemunde staff in April 1945 to Oberammergau, Bavaria, contrary to Hitler's orders.

Kammler did so to create a bargaining chip for negotiations with the OSS. Later, *Operation Sunrise*, the secret surrender of May 2, 1945 in Italy arranged between OSS operative, Allen Dulles and SS-Obergruppenführer Karl Wolff, was arranged. And even later, many of these scientists that Kammler saved were brought into the U.S. in *Operation Paperclip*, among them Dr. Wernher von Braun and Dr. Debus.

Kammler had been accused of war crimes because he was in charge at the Nordhousen underground Mittelwerk where slave laborers feverously worked to build the V-2 Rockets. Many of these laborers were worked to death and others, suspected of sabatague, were killed in mass hangings to discourage sabotage. So, Kammler had to have an identity change before starting work at Wright Patterson and worked there under a different name.

The Soviets also, either obtained another version of the Die Glocke, or obtained the scientists to construct another one. McClellan claims the mysterious object that crashed at Keckesburg, Pennsylvania was the Soviet Venera 3MV-4 Venus spacecraft, launched on November 16, 1965.

The Venus orbital insertion failed on December 9, 1965 and the spacecraft was reentering the atmosphere, heading towards the U.S. The Russians called NORAD to informed them what had happened so that they wouldn't think it was an incoming nuclear missile.

NORAD tracked the object, which not only did not burn up on reentry, but made several maneuvers to avoid heavily populated areas, to where it landed and called the proper recovery teams. McClellan was on one of those teams and inspected the object.

It suffered little or no damage and had made a long furrow in the ground before coming to a stop. The craft was still warm to his touch. McClellan knocked on the object several times to see if there was anything alive inside - with no response.

Soon, it was taken, under military escort, by a flatbed truck first to Lockbourne AFB and then to Wright Patterson where it was examined and back engineered by German scientists, which included, Dr. Hans Kammler, Dr. Kurt Debus, Rudolph F. Huelker, Author Rudolf, Dr. Ernst Geisser, Dr. Albert Zeiler and Dr. Kraft Ehricke.

All of these men also worked for NASA at the Kennedy Space Center. It was determined that the Venera 3MV-4 Venus spacecraft has highly similar to Die Glocke by these men, some of who had actually worked on Germany's original one.

Dr. Debus also revealed to McClellan that the Die Glock could also be used for time travel. The U.S. black projects have since incorporated the technology of Die Glocke into their own spacecraft, according to McClellan.

Dr. Debus also mentioned to McClellan, in passing, his belief that the Aryan race originated in the star system Aldebaran – the same place Maria Orsic was getting her channeled information to build the other world flying machine! William Tompkins stated that Dr. Debus told him he had a chance to go to Aldebaran after the war but he accepted the generous offer of the U.S. and decided to stay on this planet.

Wright Patterson was the primary location to back engineer recovered UFOs until Area 51 was established. An even more secret location, S-4 near Papoose Mountain, has hangers built into the mountain and many underground levels. When the hangar doors are closed, the camouflage is so good that you wouldn't even know they were there. S-4 is another area used for testing and back engineering recovered UFOs.

The present commander of the Air Force Research Laboratory at Wright Patterson AFB is Major General William N. McCasland. According to the Air Force's official website at: http://www.af.mil/AboutUs/Biographies/Display/tabid/225/Article/104776/major-general-william-n-mccasland.aspx

> Major General William N. McCasland is responsible for managing the Air Force's $2.2 billion science and

technology program as well as additional customer funded research and development of $2.2 billion. He is also responsible for a global workforce of approximately 10,800 people in the laboratory's component technology directorates, 711th Human Performance Wing and the Air Force Office of Scientific Research.

General McCasland was commissioned in 1979 after graduating from the U.S. Air Force Academy with a Bachelor of Science degree in astronautical engineering. He has served in a wide variety of space research, acquisition and operations roles within the Air Force and the National Reconnaissance Office. He commanded the Phillips site of Air Force Research Laboratory at Kirtland AFB, N.M., and served as Vice Commander of the Ogden Air Logistics Center and the Space and Missile Systems Center. He previously served at the Pentagon, first as the Director, Space Acquisition, in the Office of the Secretary of the Air Force, and then as Director of Special Programs, Office of the Under Secretary of Defense for Acquisition, Technology and Logistics. General McCasland holds a doctorate degree in astronautical engineering from the Massachusetts Institute of Technology where he studied under a John and Fannie Hertz Foundation fellowship.

EDUCATION

1979 Bachelor of Science degree in astronautical engineering, U.S. Air Force Academy, Colorado Springs, Colo.

1980 Master of Science degree in aeronautical engineering, Massachusetts Institute of Technology, Cambridge

1988 Doctor of Philosophy degree in astronautical engineering, Massachusetts Institute of Technology, Cambridge

1995 Air War College, Maxwell AFB, Ala.

1999 Advanced Program Manager's Course, Defense Systems Management College, Fort Belvoir, Va.

2004 United States-Russia Security Program, John F. Kennedy School of Government, Harvard University, Cambridge, Mass.

ASSIGNMENTS

1. September 1979 - September 1980, graduate student, Air Force Institute of Technology, Massachusetts Institute of Technology, Cambridge

2. October 1980 - March 1984, payload development engineer, Secretary of the Air Force Office of Special Technology and Logistics, the Pentagon, Washington, D.C.

16. May 2011 - present, Commander, Air Force Research Laboratory, Wright-Patterson AFB, Ohio Projects-6, Los Angeles AFB, Calif.

3. April 1984 - June 1985, Chief, Payload Systems Division, Secretary of the Air Force Office of Special Projects-8, Los Angeles AFB, Calif.

4. July 1985 - September 1988, doctoral student, Air Force Institute of Technology, Massachusetts Institute of

Technology, Cambridge

5. October 1988 - May 1992, assistant director, Systems Engineering, Secretary of the Air Force Office of Special Projects-13, Los Angeles AFB, Calif.

6. June 1992 - July 1994, Director, Mission Planning, Aerospace Data Facility, Buckley AFB, Colo.

7. August 1994 - May 1995, student, Air War College, Maxwell AFB, Ala.

8. June 1995 - August 1997, Commander, Operations Squadron, Aerospace Data Facility, Buckley AFB, Colo.

9. August 1997 - March 2000, Chief Engineer, Navstar GPS Joint Program Office, Los Angeles AFB, Calif.

10. April 2000 - September 2001, System Program Director, Space Based Laser Project Office, Los Angeles AFB, Calif.

11. October 2001 - May 2004, Materiel Wing Director, Air Force Research Laboratory Space Vehicles Directorate, and Commander, Phillips Research Site, Kirtland AFB, N.M.

12. June 2004 - October 2005, Vice Commander, Ogden Air Logistics Center, Hill AFB, Utah

13. October 2005 - June 2007, Vice Commander, Space and Missile Systems Center, Los Angeles AFB, Calif.

14. June 2007 - June 2009, Director, Space Acquisition,

Office of the Under Secretary of the Air Force, Washington, D.C.

15. June 2009 - May 2011, Director, Special Programs, Office of the Under Secretary of Defense for Acquisition, Technology and Logistics, the Pentagon, Washington D.C. 16. May 2011 - present, Commander, Air Force Research Laboratory, Wright-Patterson AFB, Ohio

Major General McCasland is one of the high level people quietly working from behind the scene to gradually bring full disclosure of the Secret Space Programs and the ET presence to the world's people. He was recently consulting Tom DeLonge who co-authored *Sekret Machines: Chasing Shadows,* a science fiction based on fact.

DeLonge, a former lead vocalist for Blink 182, is the creator of the Sekret Machines multimedia disclosure initiative, and has enlisted top writers such as A.J. Hartley and Peter Levenda to co-author up to six books. In the preface of *Sekret Machines* and interviews, Delonge describes how he is being helped by a team of ten advisors with direct links to corporations and Department of Defense entities involved with the U.S. development of a secret space program, and knowledge of a similar program simultaneously developed in Russia and the former Soviet Union.

Sekret Machines is the product of what the advisory team is telling Delonge about their knowledge of the secret space programs that evolved in the US and Russia. According to DeLonge, their revelations are officially sanctioned by those with need to know access, and therefore the most authoritative disclosure yet to emerge on the topics of UFOs, extraterrestrial life and secret space programs.

Another, now retired, high level person involved in disclosure is former Special Assistant to the Commander of Air Force Space Command, Major General Michael Carey who said:

> "Sekret Machines scratches at the surface of "who do" we trust with our classified technology – certainly our adversaries are aware of our undertakings, as they are doing the same, but what of our citizens, our politicians, even our own military. Tom DeLonge and A.J. Hartley create a convincing narrative describing the "cat and mouse" game that is timeless between strategic adversaries. It has existed under the sea, on the surface of the earth and in its skies, why wouldn't we believe it occurs in space. Our military leaders have been saying space is a contested environment for years now, perhaps we should believe them!"

Although quite a lot of science fiction is based on fact. I have one problem with science fiction based on fact. Where do the facts end and the fiction take over?

That is why I like to stick with nonfiction while writing books like this. Action and drama be damned. Truth *is* stranger than fiction!

James Oglesby, who worked for Bendix NASA launch support, as a clerk in the Bendix parts and tool crib department, observed a number of, what he thought to be extraterrestrial, saucer shaped craft landing in a remote field near the Kennedy Space Center in the evening hours of darkness. After tracking these landings for many months, he eventually came to the conclusion that there were clandestine meetings taking place between some NASA officials and these ETs as revealed in his books, *Proof of Extraterrestrial Intelligence: The Cape Canaveral*

Apollo Program Chronicles and *Travelers From Space*.

NASA has been quite secretive about ET encounters by Apollo Program Astronauts. But, Major Donald E. Keyhoe revealed the following in an article in *True Magazine:*

> "On April 9th, 1964, the national Aeronautics and Space Administration launched from Cape Kennedy the first two-man Gemini capsule, a crucial step in our effort to land an astronaut on the moon. The capsule went into the planned orbit around the earth, and sensitive instruments started gathering data that would show flaws and point out possible improvements in the design. This first test was a great success. You read about it the next morning in the paper. But, there was something that you didn't read.
>
> Two scientists present at the test gave this report to me confidentially. The Gemini capsule was still in its first orbit when four spaceships of unknown origin flew up to it. While startled radar trackers watched their screens in open-mouth amazement, the four took up positions around the capsule, two above it, one beneath, one aft. The occupants inside these strange ships appeared to be inspecting the capsule minutely and with care. They drew close to the capsule and paced it for a full orbit of the earth. Then, apparently finished with their scrutiny, they pulled away and vanished into the unknown.
>
> What were these four mysterious space travelers? Where had they come from? What mission had brought them into the earth's space neighborhood? What people, what beings, were at the controls? I fervently wish I could answer these questions. And, I wish I could satisfactorily

answer one other: this eerie episode, this incident so fraught with implications for all who live on earth – why was it kept secret?"

In another area of secrecy, many researchers have noted the special secret deals made between the Germans and the U.S. before and after the end of World War II. Many were made between Knights of Malta on both sides. For example, the head of the OSS was William Donovan, a Knight of Malta. The OSS hired Knight of Malta, Reinhardt Gehlen to independently run his intelligence agency of the Third Reich for U.S. intelligence after the war, Later, he was hired at CIA by another Knight of Malta, Allen Dulles.

Knights of Malta take their orders from the Jesuit General at the Vatican- the reason being as follows:

The Jesuits had been outlawed in 1773 for their many crimes and were forced underground - forming the Illuminati. The Jesuits also infiltrated and took over control of many Freemason lodges. The Illuminati hired secret representatives to persuade the Pope to hire Mayer Amshel Rothschild, a member of the Freemasons and controlled by the Illuminati, to be appointed Guardian of the Vatican treasury. After he was so appointed, the outlawed Jesuits still had access to the vast wealth of the Vatican by their secret strategy.

At the secret Masonic Congress at Whilemsbad in Meyer Rothschild's castle on July 16, 1782, the Illuminati was joined with the Freemasons and the planning for the French Revolution took place.

Later, Napoleon, secretly controlled by the Jesuits, was ordered to invade the Island of Malta, then the headquarters of the Knights of Malta. This ended up giving the Jesuits control over the Knights of Malta – the Vatican's army. Later, Napoleon was ordered to invade Italy and arrest the Pope. The Pope was informed that he would regain his freedom if he reinstated the Jesuit Order. After 5 years in prison, the Pope relented.

The Knights of Malta have been taking their orders from the Jesuit General at the Vatican since 1814, when the Jesuit Order was reinstated. Thereafter, the Jesuit General actually controlled the Vatican, the Knights of Malta and the Pope, who was reduced to a mere figurehead.

The Vatican helped many Nazi SS war criminals escape to neutral countries after the war. The Jesuits had, after all, brought them into power in the first place, as documented in The *Secret History of the New World Order* and *The Secret History of the Jesuits*. Also, there seems to be, in addition to a number of former Nazis, a number of high level Freemasons running NASA. Could they also be taking orders from the Jesuit General?

Clark McClellan, who knows that former members of the Nazi SS actually ran NASA, believes that our country, the U.S., has been taken over by the Zionists.

What I point out in my *Secret History of the New World Order*, is that the Jesuits at the Vatican created Zionism and are using the Jews as part of their plan to create a New World Order or one world government with the Capital at Jerusalem. A quick search on "Jesuit Rothschild" on Google will demonstrate part of what I discovered, and am claiming.

The Jesuit presence in the U.S. was a large factor in the Nazis gaining secret control of power in the U.S. at not only at NASA, but at the CIA, which was created by Knights of Malta, Allen

Dulles and William Donovan, and the Military Industrial Complex as well.

William Tomkins had several interviews with von Braun and Dr. Debus at the Kennedy Space Center (KSC) and noted that the older man, Debus acted like the boss of von Braun on these occasions. This seemed strange because at the time, von Braun was the manager at KSC and outranked Debus.

However, it must be remembered that Debus was the Manager at the V-2 rocket launching facility at Peenemunde in Germany during World War II, and there, he outranked von Braun, who was his employee there. It could be that secretly; Dr. Debus was the real man in charge at NASA while the charming Von Braun was, as a public relations man, used to deflect attention away from Debus.

As William Tomkins states, he was working on the Apollo moon landing project at the Advanced Design department of Douglas 4 years before NASA was even created. Their planning called for a 2,000-man Navy moon base and many more moon missions than actually ended up happening. About 400,000 people were directly working on the Apollo program. The Apollo Command module was testing C3I communications systems that were planned to be used on future Navy space ships.

During the final two years of his 12-year employment at the Douglas Aviation Company, Tompkins' innovative designs for planned Apollo missions had greatly impressed Dr. Kurt H. Debus. In July 1962, Debus had become the first Director of NASA's Launch Operation Center (renamed the Kennedy Space Center after the JFK Assassination), a position he held until his retirement in November 1974.

In 1963, Debus appointed Tompkins to a Working Group for the future Launch Operations Center. The two held many confidential meetings over the future of the Apollo program, and discussed its real mission as part of an ambitious Navy space program called "NOVA" for secretly establishing military garrisons on the Moon, Mars and nearby star systems.

The Apollo Moon landings were only the first stage of an ambitious four stage plan for NOVA. Stage 2 of NOVA was to put 10,000 people on the Moon. Stage 3 was to place bases on Mars and other planetary bodies in the Solar System. Finally, Stage 4 was to place manned Navy bases in 12 adjoining star systems.

All of this was going according to plan right up to 1968 when a spanner wrench was thrown into all their plans.

The Manned Apollo 8 orbiting lunar Command Service Module (CSM) spotted extraterrestrial space ships closely crossing their orbital path around the moon. Walter Cunningham radioed to Houston Control center, "There is a Santa Claus and a Mrs. Claus too." Houston replied, "You were instructed not to discuss that."

"Santa Claus" was a NASA code word for UFO. Other frequent code words used were "Barbara" for artifacts or any ruins and "go to KILO" meaning make communications less obvious. Commander Shira said "Look at the size of those alien ships! They are coming in fast at 2 O'clock, straight at us." At the time, he was probably too frightened to stick with NASA protocol.

They were debating changing orbit to avoid a collision with the huge oncoming space ship. Then, Cunningham said "No they will pass in front of us" It was decided not to change orbit to avoid the oncoming space ship which narrowly missed the Apollo 8 CSM.

These ET craft were estimated to be about 1.5 Kilometers long by the people at the command center. The Apollo 8 CSM also returned photographs of extraterrestrial bases on the lunar surface. These photos were quickly classified by the CIA as was all reference to ETs or UFOs in the astronaut's messages.

Even so, it was decided to proceed with the manned Moon landing and see what would happen. In 1966 the lunar Orbiter 2, had passed over the Sea of Tranquility at an altitude of 30 miles. The onboard camera photographed 6 pyramids arranged in geometrical patterns. Three were aligned like the Pyramids at Giza, patterning the three stars in the belt of Orion. It was decided the first lunar landing would be in this area.

When The Apollo LEM was preparing to land on the Moon, the crew sent the message, "Oh my God, you wouldn't believe it. These babies are huge, sir, enormous. I'm telling you there are other spacecraft out there, lined up on the far side of the crater edge, sir. They are on the Moon watching us."

While this part of the transmission was cut off from the public, some Ham radio operators will attest that that is what they heard on their microwave receivers, while listening to the Lunar landing. There were a number of tremendous sized ET space craft parked on the far side of the crater watching them! The covert Navy NOVA plan came to a crushing end during the Apollo 11 mission when Neil Armstrong and Buzz Aldrin were met by a fleet of menacing extraterrestrial spacecraft.

During the Apollo 11 Moon landing in July 1969, Tompkins says he was in the NASA Launch Operations Center at Cape

Canaveral as part of a large TRW contingent. He states that television cameras from the Apollo Lander provided a live feed of what was being witnessed by Armstrong and Aldrin.

William Tompkins was at the TRW monitoring station with a live video feed from the LEM. When Armstrong was photographing Buzz Aldrin's space helmet, Tompkins and others present at TRW could see actual ruins of a former civilization on the lunar surface, reflected in his helmet. Then, they heard an alien voice, "finish a total of six of your Apollo missions; take your photos, pick up some rocks, go home and don't come back."

The extraterrestrial occupants of the large starships intimidating the Apollo 11 mission did not want the U.S. Navy establishing a beachhead for future military bases on the Moon. By preventing the U.S. Navy in moving forward with its plan to put 10,000 people on the Moon using a number of NOVA rocket launchers throughout the 1970's, the NOVA program effectively came to a crushing end in July 1969.

It would take the U.S. Navy more than a decade before it could complete the construction of its first antigravity space vehicles as part of a program called Solar Warden. According to Tompkins and other whistleblowers, the first U.S. Navy space battle groups were deployed in the early 1980's during the Reagan Administration, thereby establishing a U.S. Navy presence in deep space for the first time.

The no trespassing sign had gone up! And, that put an end to Apollo after 1972. And, that is why NASA never went back to the Moon. About 400,000 people were laid off at Douglas, Boeing, Grumman, North American, ITT, Caltech, JPL, and numerous other companies involved with the Apollo Program.

Hyper Dimensional Physics

The mathematical and physical parameters required to describe higher dimensions were primarily founded in the pioneering work of several 19th Century founders of modern mathematics and physics. Among these were German mathematician, Georg Riemann; Scottish physicist, Sir William Thompson; Scottish physicist, James Clerk Maxwell; and British Mathematician, Sir William Rowan Hamilton.

In 1867, Thompson, following decades of inquiry into the fundamental properties of both matter and the space between, proposed a radical new explanation for the most fundamental properties of solid objects — the existence of the vortex atom composed of vortexes in the all-pervading ether. This was in direct contradiction to then prevailing 19th Century theories of matter, in which atoms were still viewed as infinitesimal "small, hard bodies imagined by Lucretius, and endorsed by Newton.

Even as Thompson published his revolutionary model for the atom, Maxwell, building on Thompson's earlier explorations of the underlying properties of this etheric fluid, was well on the way to devising a highly successful mechanical vortex model of the incompressible ether itself, in which Thompson's vortex atom could live — created a model derived in part from the laboratory-observed elastic and dynamical properties of solids.

Georg Bernard Riemann mathematically initiated the 19th Century scientific community into the idea of "hyperspace," on June 10, 1854. In a seminal presentation made at the University of Gottinggen in Germany, Riemann put forth the first mathematical description of the possibility of "higher, unseen dimensions ..." under the deceptively simple title: *"On the Hypotheses Which Lie at the Foundation of Geometry."*

Riemann's paper was a fundamental assault on the 2000-year old assumptions of "Euclidian Geometry" — the ordered, rectilinear laws of "ordinary" three dimensional reality. In its place, Riemann proposed a four-dimensional reality (of which our 3-D reality was merely a "subset"), in which the geometric rules were radically different, but also internally self-consistent.

Even more radical: Riemann proposed that the basic laws of nature in 3-space, the three mysterious forces then known to physics — electrostatics, magnetism and gravity — were all fundamentally united in 4-space, and merely "looked different" because of the resulting "crumpled geometry" of our three-dimensional reality.

In terms of actual physics, Riemann was suggesting something clearly revolutionary: a major break with Newton's "force creates action-at-a-distance" theories of the time, which had been proposed to explain the "magical" properties of magnetic and electrical attraction and repulsion, gravitationally-curved motions of planets ... and falling apples, for over 200 years; in place of Newton, Riemann was proposing that such "apparent forces'" are a direct result of objects moving through 3-space "geometry" ... distorted by the intruding geometry of "4-space!"

Ultimately, in 1873, Maxwell would succeed in uniting a couple hundred years of electrical and magnetic scientific observations into a comprehensive, overarching electromagnetic theory of light vibrations, carried across space by this "incompressible and highly stressed universal etheric fluid.

Maxwell's mathematical basis for his triumphant unification of these two great mystery forces of 19th Century physics were quaternions - a term adopted in the 1840s by mathematician Sir William Rowan Hamilton, for "an ordered pair of complex numbers" (quaternion = four).

Complex numbers themselves, according to Hamilton's clarifications of long-mysterious terms such as "imaginary" and "real" numbers utilized in earlier definitions, were nothing more than "pairs of real numbers which are added or multiplied according to certain formal rules." In 1897, A.S. Hathaway formally extended Hamilton's ideas regarding quaternions as "sets of four real numbers" to the idea of four spatial dimensions, in a paper entitled "*Quaternions as numbers of four-dimensional space,*" published in the Bulletin of the American Mathematical Society [4 (1887), 54-7].

It is obvious from Maxwell's own writings that, even before Hathaway's formalization, his choice of quaternions as mathematical operators for his electromagnetic theory was based on his belief that three-dimensional physical phenomena (including even perhaps the basis of human consciousness itself) are dependent upon higher dimensional realities.

So in effect Maxwell's original equations were based on quaternions or 4 dimensional numbers. The tragedy is that the work of this brilliant theoretician, was quickly modified by men of less intellectual capability.

One was Oliver Heaviside, who was described by Scientific American (Sept. 1950) as "self-taught and ... never connected with any university." Heaviside actually felt that Maxwell's use of quaternions and their description of the "potentials" of space was mystical and should be removed from the theory. So, he eliminated the scalar components of the quaternions and vectorized the remaining portions. In this process, he eliminated the hyperspatial components of these quaternions.

Heaviside thought that potentials were mystical because there was no way to measure them. Only potential difference can be measured - but not absolute potential. Where he was wrong was the fact, now proven in quantum mechanics, that potentials are more causal than fields. For example, the Magnetic Vector Potential, A, generates the Electric field by taking its derivative with respect to time and the Magnetic field by taking its Vector Curl.

This limited subset of Maxwell's equations is what is taught in the Universities – not the real Maxwell Equations!

One scientist to point out the ramifications of the truncation of Maxwell's original theory is Thomas Bearden who has written a number of books and papers on this subject. Bearden has for example proposed a solution to the source charge problem which uses hyperspace principles.

The question "what is electric charge?" is still not satisfactorily answered in classical physics.

To model the source charge problem in easy to understand concepts, let's suppose that the electric field is a stream of moving ether flowing out of the electrons and flowing into the protons. The source charge problem is then "where does the ether come from in the electron and where does it go in the proton?" In physics this is called a broken 3 symmetry.

Another mystery is that in quantum mechanics, one complete spin of the electron or proton take place in 720 degrees – not 360 degrees! This is explained that the spin is hyper spatial with 360 degrees of spin in real (xyz) space and 360 degrees of spin in imaginary space (-ict) or time.

Bearden's solution is that the EM energy (ether) flows into the electron during the imaginary, (-ict) 360-degree portion of its spin and gives it out during the real space (xyz) 360-degree part of the spin. Conversely, the proton receives the energy during the real (xyz) part of its spin and gives it out during the imaginary (-ict) part of the spin.

So, the ether (or EM energy) flow is conserved because it flows into the electron along the time dimension and out along the space dimensions towards the proton which takes it in and then flows it out along the time dimension. The hyperspace (or 4 dimensional) spin factor seems to be a mechanism in creating this dipole effect. And 4 dimensions are needed to keep energy or ether flow conservation between electron and proton of this dipole.

Thomas Bearden has also conceptually proposed an engineerable unified EM-GR theory. He has shown that Maxwell's original theory using 4 dimensional numbers, probably unknown to Maxwell, was actually a unified theory connecting Gravity to Electromagnetism. Here is what Bearden has to say in his paper *"Possible Whittaker Unification of Electromagnetics, General Relativity, and Quantum Mechanics,"* to the Association of Distinguished American Scientists 2311 Big Cove Road, Huntsville, Alabama, 35801:

" ... In discarding the scalar component of the quaternion, Heaviside and Gibbs unwittingly discarded the unified EM/G [electromagnetic/ gravitational] portion of Maxwell's theory that arises when the translation/ directional components of two interacting quaternions reduce to zero, but the scalar resultant remains and infolds a deterministic, dynamic structure that is a function of oppositive directional/translational components. In the infolding of EM energy inside a scalar potential, a structured scalar potential results, almost precisely as later shown by Whittaker but unnoticed by the scientific community. The simple vector equations produced by Heaviside and Gibbs captured only that subset of Maxwell's theory where EM and gravitation are mutually exclusive. In that subset, electromagnetic circuits and equipment will not ever, and cannot ever, produce gravitational or inertial effects in materials and equipment.

Brutally, not a single one of those Heaviside/Gibbs equations ever appeared in a paper or book by James Clerk Maxwell, even though the severely restricted Heaviside/ Gibbs interpretation is universally and erroneously taught in all Western universities as Maxwell's theory.

As a result of this artificial restriction of Maxwell's theory, Einstein also inadvertently restricted his theory of general relativity, forever preventing the unification of electromagnetics and relativity. He also essentially prevented the present restricted general relativity from ever becoming an experimental, engineerable science on the laboratory bench, since a hidden internalized electromagnetics causing a deterministically structured local space-time curvature was excluded.

Quantum mechanics used only the Heaviside/Gibbs externalized electromagnetics and completely missed Maxwell's internalized and ordered electromagnetics enfolded inside a structured scalar potential. Accordingly, QM [quantum mechanics] maintained its Gibbs statistics of quantum change, which is not chaotic a priori. Quantum physicists by and large excluded Bohm's hidden variable theory, which conceivably could have offered the potential of engineering quantum change — engineering physical reality itself.

Each of these major scientific disciplines missed and excluded a subset of their disciplinary area, because they did not have the scalar component of the quaternion to incorporate. Further, they completely missed the significance of the Whittaker approach, which already shows how to apply and engineer the very subsets they had excluded.

What now exists in these areas are three separate, inconsistent disciplines. Each of them unwittingly excluded a vital part of its discipline, which was the unified field part. Ironically, then, present physicists continue to exert great effort to find the missing key to unification of the three disciplines, but find it hopeless, because these special subsets are already contradictory to one another, as is quite well-known to foundations physicists.

Obviously, if one wishes to unify physics, one must add back the unintentionally excluded, unifying subsets to each discipline. Interestingly, all three needed subsets turn out to be one and the same ..."

Bearden also has pointed out that electric fields can warp

space-time vastly more efficiently than mass and the controlled usage of powerful space-time curvatures can be used for anti-gravity and unilateral space thrust systems. This has been well proven out by the experiments of Thomas Townsend Brown since the 1920s. (5)

In my book, *The Covert Colonization of our Solar System,* in the chapter on scalar physics, I demonstrate the math theory that gravitational potential is proportional to the vector divergence of the Magnetic Vector Potential, **A**. Thus showing, how to engineer artificial gravity (or anti-gravity) electromagnetically.

One of the least understood concepts known to man is that of *time*. A great deal of headway regarding the nature of time was made by engineer Dewey B. Larson, published in his 1959 book, *The Structure of the Physical Universe*. Larson asserts that both space and time are simply the *aspects* of a reciprocal ratio that he refers to as *motion*, have no other meaning, and cannot exist independently outside of this relation.

He posits that everything is motion which is expressed as distance divided by time, yielding the reciprocal relation between space and time. The concept of space-time is analogous to the Eastern concepts of yin-yang, where space is the *yang* aspect, and time the *yin*. As discussed in taijitu symbolism, yin-yang cannot be separated, just as Larson's ratio of space to time cannot be separated.

Larson took it one step further, observing that all the characteristics of space must also have a similar character in time. In space, we see a 3-dimensional, coordinate spatial grid with clock time. From the reciprocal perspective, there must also exist a realm that contains *3-dimensional, coordinate time* with *clock space*. He refers to the former as the "material sector" and the latter as the "cosmic sector"

The material sector is our common reference frame, that contains the observable and measurable structures of the universe. The cosmic sector, however, remains unobservable and unmeasurable to our physical senses, though we can see its effects on how *time changes space*, those effects being called *force fields* (electric, magnetic and gravitational fields). We cannot see a magnetic line of force until it interacts with a material object such as iron filings, and alters their behavior in space.

The material and cosmic sectors are better described as *complex conjugates* of each other, so *space is real* and *time is imaginary*—but not in the sense of "make believe," but in the sense of the imaginary number. Understand the *complex number*, a combination of real and imaginary, and you'll understand the connection between space and time; yang and yin; body and soul.

Since you know that coordinate time isn't imaginary, those funny little imaginary numbers are actually showing an interaction between the physical and metaphysical... and with a foot in both realms, it does open the door to those things that are beyond space and time.

The Reciprocal Space Time theory has been very successful in describing many properties of matter and fields. A Website dedicated to Larson's work is here: http://reciprocalsystem.org/

This work and has been even more improved by others who have taken Larson's work even further over many decades of work. This website gives a detailed description of this work: http://rs2theory.org

In Einstein's relativity, the time coordinate is expressed as -ict. Here i is the imaginary number or the square root of -1, c is the speed of light and t is time. In the complex plane real and imaginary numbers are combined in an orthogonal relationship. Space and time also have this orthogonal relationship as well as the reciprocal relationship. Under normal circumstances, there is zero projection of time onto space and zero projection of space onto time.

However, as we shall see, under non-normal circumstances this may not always be true and herein lies the secret of teleportation, time travel and faster than light travel.

Recovered ET Space Craft

Obviously, if a ET crashed spacecraft could be recovered and its technical hardware studied and back-engineered into reproducible hardware, it would highly advantageous to the scientists and engineers developing the secret space program. So, that is exactly what the people of this highly secret project decided to do.

During World War II, there were many sightings of "Foo Fighters", mysterious glowing objects that flew close to bombers and fighter planes. Some in the Army Air Force thought they were German secret weapons being tested and others thought they could be of extraterrestrial origin.

A top secret memo, dated March 5, 1942, addressed to president Roosevelt, explained how the Navy had recovered an "unidentified airplane...of interplanetary origin". This craft was recovered off of San Diego shortly after the famous Battle of Los Angles. It may or may not have been related to that event.

In any case, because of these events, under the direction of General McArthur, Army Chief of Staff, General George Marshall ordered the crashed saucer studied and created the highly classified Interplanetary Phenomena Unit (IPU), within the Army's G-2 Intelligence Agency.

In the beginning in June of 1947, the, U.S. Army Counter Intelligence Corps (CIC) was in charge of investigating UFOs. But, after the Air Force became separate from the Army because of the National Security Act, on September 23, 1947, the CIC officers investigating UFOs went over to the Air Force Office of Special Investigations (OSI).

To serve as a clearinghouse for all the UFO reports coming in, the Air Force created Project Sign on January 22, 1948. In February 22, 1949, the Air Force claimed that they had shut down Project Sign and released a false report for public consumption, stating that all UFOs could be explained away by hoaxes, illusions, misidentification of known objects or aircraft observed under unusual conditions.

In reality, Project Sign was merely renamed Project Grudge and continued analyzing UFO reports until 1952, when it was again renamed Project Blue Book until its demise in December of 1969, at which time the U.S. government stated that they would no longer investigate reports of UFOs.

This was the public record, designed to place the public at ease concerning UFOs. However, the real UFO research was being done in secret at Wright Patterson AFB in Dayton Ohio.

The North American Aerospace Defense Command (NORAD) is a United States and Canada bi-national organization charged with the missions of aerospace warning and aerospace control for North America. Aerospace warning includes the detection, validation, and warning of attack against North America whether by aircraft, missiles, or space vehicles, through mutual support arrangements with other commands.

NORAD would signal special "Delta" recovery teams whenever they tracked a UFO which appeared to crash and direct that Delta team towards the estimated crash site.

This would also be done for some non U.S. UFO crash sites, where special secret treaties would allow these Delta teams to enter other countries and recover crashed UFOs.

The 4602nd Aircraft squadron had a World War II mission to recover downed enemy air craft and do field exploitation of enemy hardware, personnel, and documentation. On January of 1953, the Air Force created the 4602 nd Air Intelligence Service Squad (AISS). This unit was used in peacetime to do field exploitation of downed UFOs. They also created Operation Blue Fly and Operation Moon Dust, which covered these UFO field exploitations. Also, there are pro words used like; "Round Robin", that signifies a spacecraft of another earthly government or "Global Horizon", that signifies a spacecraft not from the Earth. For example, Round Robin would apply to recovered German flying saucers.

During the war, Wright Patterson had a section called Air Material Command. Captured enemy advanced aircraft were sent there for examination and back-engineering. So, it was only natural that later recovered crashed UFOs would be also sent there. Air material Command was later renamed Air Tactical Intelligence Center (ATIC).

In the early days, the Air Force also had a Detachment 35, which collected data, pictures, reports, photographs and physical evidence dealing with the more reliable reports of UFOs. The evidence of the Roswell UFO crash was handed over to

Detachment 35 by U.S. Army CIC.

During crashed UFO recovery operations, often live ETs would be recovered, which the military teams would call Extraterrestrial Biological Entities or EBEs. Even more valuable to the military than the ET hardware was information obtained from the surviving live EBEs. They could inform the military what their mission on Earth was, how many other ET groups were visiting the Earth and what their missions were, how did their technology work and so much more.

But to receive this valuable information, a means of communication with the EBEs was required. There are some persons who, particularly at a young age before puberty, have advanced phychic abilities, including telepathy. A means of testing for these abilities was created and a test was given in the public schools at the kindergarten and first grade level to discover any of these rare and gifted children. This test seemed innocent enough and the teachers giving these tests had no idea what their real purpose was. In any case, those heading these secret military programs were keeping track of the occasional child with advanced intuitive abilities.

Some of these children were secretly inducted into MILAB programs like Corey Goode, an identified Intuitive Empath at age 6. MILAB stands for military abduction. Techniques like time travel and teleportation are often used to abduct the child while in bed at night, using teleportation to undergo training for hours and to return the child back to bed to the time of the original abduction using time travel.

This way the training would not be known about by the parents, who would not even realize that their child had been missing for hours. If the child would mention what he or she had experienced during training, the parents would usually think that their child just had an overactive imagination. Before long, the child would learn not to mention their experience at all.

Another MILAB child of record was Randy Cramer, who was not so intuitive, but trained to be a super soldier. He was then inducted into the Mars Defense Force after his training was completed.

Clifford Stone is unique among those experiencing and investigating UFO and ET phenomenon. He spent 22 years in the US Army as a part of an extremely elite and secret group that was rapidly dispatched to crash sites in order to recover UFO or ET craft, bodies, and artifacts. Since his retirement from the Army, he has devoted his time to a diligent Freedom of Information Act search of government archives.

He maintains that we have knowledge that intelligent life is visiting this planet in craft capable of traveling distances of many light years very quickly; effectively bypassing acceptably known physics. Further, he stresses that our recovery of these ET craft and artifacts have allowed our government to make staggering scientific gains of great potential benefit to the world.

He maintains, as do many others, that this information is held in deeply secret programs beyond Constitutional controls and safeguards, and that despite the end of the cold war, those controlling these 'black projects' have continued to keep these important discoveries to themselves and for motives known only to them.

Clifford Stone was a member of the 4602nd Air Intelligence Service Squad (AISS), located at Ft. Belvoir, Virginia that recovered crashed UFOs.

Stone says he was initially given training at a Nuclear, Biological and Chemical Warfare facility at Fort Ft. McClellan, Alabama, and then given regular army assignments until called away to perform his crash retrieval duties when required. Stone claims that when required for UFO retrievals he was typically called out to serve for a week. He says that his army service record was distorted to exclude references to his actual training and assignments, and refers to him only performing clerical duties as a typist. Like many whistleblowers, there is controversy over inconsistencies between his testimony and lack of documentary evidence to verify his alleged training and actual service.

The strongest objective support for his claims of having worked in covert UFO retrieval projects is the extensive documentation he has provided to support the existence of covert UFO crash retrieval teams. He was able to use very specific Freedom of Information Acts requests to uncover information disclosing the existence of classified projects such as Moon Dust and Blue Fly that were created to recover debris of UFOs.

Clifford Stone had a different story than most children as a child. He was interacting with ETs as a child. He actually befriended one he called Korona, who ended up being like a guardian Angel to Clifford.

Korona could make himself invisible to others while Clifford Stone could still see him. Clifford was also discovered by the military at an early age and was also being tracked by them. At a local drug store, a Captain Brown befriended young Clifford, who was around the age of seven at the time. Captain Brown had a talk with Clifford's parents. What they talked about Clifford had no idea.

And then they went ahead and said, "Okay", cause Captain

Brown says, "Well, we're friends now, so I'm going to be coming back to see him from time to time."

And apparently, Clifford's family agreed to this. And from that time on, every couple weeks or so, he would run into Captain Brown. And Captain Brown would want to know if anything new has come up.

And Clifford kept him appraised of the situation. That went on until about March of 1968 when Captain Brown was killed in a crash of a T33 out of Laughlin Air Force Base.

In 1967, a military recruiter had shown up at Clifford Stone's High School. Clifford scored well on all the military written tests but failed the physical test. So, Clifford was rated 4-F. However, after graduation, Clifford got another letter from the draft board saying that they wanted to reevaluate him.

Again he scored well on all the written tests and was selected for officer's school if he wanted to go. He still had medical problems and a Captain came out and informed Clifford that they would be sending him home soon. Then, a Colonel came out and sent the Captain home! The Colonel verified to his satisfaction that Clifford really wanted to be in the military and then said that they would give him a chance. Clifford Stone joined the Army at age nineteen.

He was tested for his job in a most unusual way. He was invited to visit the Pentagon by a military officer acquaintance of his, named Jack. But when he arrived at the Pentagon he was instead met by a friend of Jack who said Jack was called away on an important mission and that he would take Clifford on the Pentagon tour. He gave Clifford an ID card and inside they went.

Their first stop was a sort of museum room. He was shown newspaper reports and photos of the 1952 UFO flyover of Washington D.C. Here is how it went in Clifford's own words:

> "And he says, 'You're aware, of course, that 68 UFOs were picked up on the night of August 18, 1952?'
>
> And I said, "Oh, yeah. I'm very aware of that."
>
> Then he says, "Well, you know, the most unique case, even though all these get all the publicity, was the night of July 19 and 20. That was the most unique one. Most people don't know anything about it. And he carried on the conversation, bits and pieces about it.
>
> Then we get to an elevator. And he says, "Well, I'm going to show you the basement of this place." He says, "People haven't seen this, and, of course, we have to harden", which means that you prepare a building for a nuclear strike, "so we have to harden the Pentagon to make sure that people survive in the event of a nuclear attack."
>
> So he took me on down. And we got down . . . I don't know how many floors down.
>
> But we got out and there was this little silver car. I mean, you couldn't tell which was the front, and which was the back other than where the seats were, and the seats were facing one direction.
>
> And we got into that, looked like a little bullet. We got into it, and he says, "This is called a monorail." He says, "It's not on a track." And he showed me where you have

a little tube-like, and it drove on that. And it was electromatically driven.

So he went ahead, and we got in there. And I don't know how long we drove underneath there, but he was trying to tell me that the Pentagon's a mighty big place. So don't be concerned that we're driving for . . . in that little con . . . and there's no driver. You know where you're going and . . . I'm sure he had some way to control it, but I don't remember all of that about it.

I was just taken aback and fascinated, because this was the first time I'd seen anything like that.

But we got to this one place and there was this door on the side. We get out and we go into that door and there's this long corridor – no doors, just a long corridor. And this is underneath the, allegedly, underneath the Pentagon.

And I know we drove for at least 20 minutes.

But we went down that corridor, and he was telling me, "You know, a lot of things aren't the way they seem."

He says, "This looks just like it's a long corridor. You walk down to the end, there's nothing there. You have to turn around and come back. You can see the door at the other end."

And I said, "Yeah. What's your point?"

"Well, a lot of things aren't as they seem to be."

He tapped on the wall, and he says, "Solid wall, right?"

And I said, "Yeah." And I started to say, "What's your point?" Before I could say anything, he says, "It isn't necessarily solid." And he pushed me. And I go through the wall.

You know, there's nothing there, but yet, when I was there, it looked like it was just a solid wall.

And I go, "What the heck are you doing?" But before I could get up and say, "What the heck are you doing?" I notice I'm in a room. I turn around. When I turn around, there is this, what we call a field desk, which is nothing more than a little table.

And setting at the field desk, is what you would call your typical Grey.

And, once again – people get upset with it – I'm going to have to say he was about 4 ½, maybe even 5-foot tall. But he's sitting there, and he's got his hands up like this [on desk], and he's looking directly at me.

On either side of him . . . and I'm not going to say black suit, it's a dark suit, and they had dark glasses on. And they were standing there like this. [Standing up straight.] Not one word was said.

And I was the one that when I got up and turned around to see, "What the heck are you doing?" That's what I said. I remember to this day. "What the heck you do . . . ?"

And I stopped right there because I saw this, and

immediately it's like a buzz saw going off in my head. I went down on my knees, and I went face down first. I remember that. That's the last thing I remember.

I wake up. I'm back in Jack's office. I'm being told nothing happened. I must have dreamed it. No one took me any place. We had been there all along, and I seemed to be tired. And I must have just dozed off."

According to Corey Goode, who was part of the interview from which the above dialogue was excerpted, this was typical of the kind of tests that Intuitive Empaths would be put through to see if they could psychologically hold up in actual meetings with ETs. He added that some Greys can reach into your mind and cause what Clifford experienced. Apparently, Clifford passed this unusual kind of test. (6)

Clifford Stone's story about underground facilities beneath the Pentagon are supported by William Tompkins interview on Rense Radio, where Tompkins speaks of a complete underground city beneath the pentagon called *Crystal City*. That interview can be heard at http://Rense.com under William Tompkins interviews 10-31-16, hour 1.

Then, he was tested again. His first job was to approach a downed UFO near Indiantown Gap in Lebanon County, Pennsylvania. He was told that it was a Soviet craft that had crashed. He was told that they were going to recover the craft and they needed to check for radiation to insure it could be recovered safely. He was ordered to pick up his APD27 Geiger Counter and to go alone towards the craft and take readings and shout out the readings every so many feet.

When he got close to the craft he observed a dead, Grey ET

hanging out of a hatch. He yelled out, "I need an officer here, I need an officer up there." The officer in charge yelled back, "you don't need no one up there, no one is coming up, just go ahead and give us your readings, what do you see son?"

Clifford replied, "Well, what I'm seeing . . . it's not from this world, and you guys know that." I said, "Why the hell are we lying? Why are we telling people that these things aren't happening?

What do you want me to do now? I need an officer up here."

The Colonel replied, "It's okay, son. Come on back down."

Having done his job, he was removed from the scene and the craft was later recovered. He found out later that all the Grey ETs in the craft were dead.

In other cases, the ETs survived the crash of their space craft and Clifford Stone used his abilities to interface or telepathically communicate with them, often for long periods of time

Clifford Stone, said the following from information and documents he had received from others in his book *Eyes Only: The Story of Clifford Stone and UFO Crash Retrievals*

> "What happened in July 1947 here in Roswell was in fact the crash of two or more unidentified flying objects... We positively confirmed that they were interplanetary craft from some other place in our galaxy. They did not originate in our solar system. They were also bodies. We recovered the debris, we recovered whole craft. They have reverence for their dead. I can tell you that we did not maintain for too awful long, possession of the

bodies. In one instance, there were five entities. Three were dead, one was seriously injured and died later as a result of its injuries. One wasn't all that bad as far as its injuries were concerned. I cannot confirm whether it was returned dead or alive. "

Clifford then added from his own experience:

I do know that we had a problem because medication as we understand it doesn't work for them. Food that we eat, that may be healthy for us, was poisonous for them. So synthetics had to be applied, and how they went about this, I don't know, because I wasn't in the position to have any knowledge of that. What I am saying are things that I know for a fact to be true simply because I lived with them and I worked with them...."

Clifford Stone was one of many who came forward to testify in the Disclosure Conference in 2001. He claims that the military has catalogued 57 different varieties of Extraterrestrials complete with medical procedures to be used with each type.

Stone first saw the "EBE Guidebook" in 1979 and claims that it contained much information on each of group of EBEs in terms of their physiology, food requirements and medical information. He claims he could read the Guidebook when he was serving on the retrieval teams up until 1989. Stone says that the Guidebook was to be used in case First Aid had to be administered to any EBEs found at crash sites. In the interview, he gave the example of Iodine which can be administered for first aid purposes, but can be deadly for some EBEs

Telepathy is often used in communicating with the ETs but some can speak perfectly good English or the language of the country they appear in. Some ETs even have translator devices that can translate between their language and ours.

In another incident, Clifford Stone was taken to a wooded area surrounding an Air Base, he thought was Bowling AFB. He was briefed that a UFO had crashed nearby and a live EBE had been taken to the Base. When he was taken into the room where the grayish white entity was held, the first thought message he heard was "I am afraid."

Clifford telepathically responded with "Are you not a guest here, are you not here of your own free will?"

The entity responded with, "I am a prisoner, I'm captured. And, I am afraid". The entity was obviously telling the truth. Both his hands were chained to restraints where he was seated even while the officer in charge was telling Clifford that he was a guest.

Clifford returned, "Well what about your people? Will they be coming to get you?"

The entity responded with, "They were going to pick me up where I was at, before your people arrived."

Clifford could also empathically feel the entity's emotions which overwhelmed him with compassion for this being. He was required to verbalize the telepathic communication, while a stenographer in the room would quickly write down their conversation. It was then, that he decided to help the ET escape back to his own people.

Further telepathic communication was partially kept secret between the two. Clifford remembered that he had seen a friend of his on the recovery team while entering the base.

Clifford stated that he needed a break outside for a while. The officer in charge, a Colonel agreed. Outside he asked his friend to leave a pair of wire cutters hidden in a specified place outside the building where the ET interrogations were being held. Then he returned inside.

He confessed that part of their telepathic conversation had been kept secret. The Colonel responded, "O.K., you want to tell me what it is now?"

"Better still Sir, I need to get everyone cleared from this building. I need you to go ahead, get somewhere around a hundred yards away. If you will give me the keys to his handcuffs...because I may have to free one hand".

The Colonel was at first hesitant. So, Clifford continued, "We are going to see something very spectacular here...but I have to be here to assist this entity."

The Colonel responded, "If we pull back, will you be in any danger?" Clifford responded, "I don't think so but I am willing to accept that risk. This may be something very important for all of us."

The men pulled back. Clifford asked the entity," How far do you need to be before something could happen." The entity replied, "They are close by now. If I get 10 to 15 feet outside this perimeter, I will be able to survive and be picked up by my kind."

Clifford picked up the PRC 9 communication device and explained to the men who had pulled back, "O.K. we are going to start the sequence of events at this time. Watch the skies". Then he freed the captive and they quickly headed outside. A hole was cut in the perimeter wire and the ET slipped through. Soon, a brilliantly lit UFO appeared overhead. And quickly swooped down and picked up the freed captive.

The Colonel soon thereafter angrily confronted Clifford. He was asked if he was familiar with the uniform code of military justice as it applies to helping an enemy to escape. Clifford responded, "Yes Sir, I was. But, I was told that this individual is a guest." The Colonel responded, "Well he is, but we needed to get some vital technical information from our visitor". Clifford Stone was not court-marshalled. He was too valuable to their interfacing program.

The number of recovered crashed UFOs is quite numerous and number in the hundreds worldwide. Because of the high level of secrecy, only a few are known about to most researchers and even Clifford Stone doesn't know about them all.

Also, there are cases where ETs are living among us - usually without our knowledge. Clifford Stone mentions a ET family living in Texas that actually had a flying saucer parked in their barn.

That is, until they were discovered and had to relocate to another State. Many of the contactees of the 1950s described the UFO personnel as looking quite human, like they may have come from Sweden or Norway.

There is the classic case of Valiant Thor, who actually lived at the Pentagon for several years as documented in *The Stranger at the Pentagon* by Dr. Frank E. Stranges. Valiant Thor, was from Venus. The Venusians had been monitoring the Earth for centuries, but after the event of atomic bombs, these missions increased dramatically.

Valiant Thor first landed his craft near High Bridge, New Jersey to join a gathering of UFO contactees, headed by Howard Menger. Several photos of Menger and Valiant Thor, seen together in High Bridge New Jersey, exist. They are shown in the book, *The Stranger at the Pentagon*.

After that meeting, Valiant Thor landed his craft in the outskirts of Arlington, Virginia. He was arrested by a local police officer. After some radio conversation on his two-way radio, the policeman took Valiant Thor to the Pentagon. There, he introduced himself at the Pentagon, where he stayed as a guest for three and a half years, trying to discover why mankind on Earth was so possessed with trying to destroy each other in senseless and insane wars. He also offered advice, when asked.

Howard Menger actually claims to have been taken to both the Venus and the Moon by ETs, as described in his book *From Outer Space To You*. This book is also recommended reading for the spiritual teachings of the beings from other planets and star systems and the trials and experiences of Earth based UFO contactees as well as the technical descriptions of ET hardware.

The Trip to Venus was a flyby without actually landing. Through the viewing portholes, Menger could see the buildings, parks, people and unusual animals of Venus. The whole round trip from Earth to Venus and back, he estimated to take about a half of one hour.

His first trip to the Moon was also a fly by, at a higher altitude and also lasted about half an hour. His second trip to the Moon was on a larger saucer and other UFO contactees were also brought along, some of whom he recognized. This time, they actually landed on a base on the moon. There they met with other contactees from other countries of Earth - and their Moon based tour guides.

The landscape of the Moon reminded Howard of the Arizona deserts, hot and dry. At one point the passengers left the protective environment of their levitating tour bus and Howard discovered that he could actually breathe the lunar air! However, the outside temperature was so hot, that he was glad to get back inside the air-conditioned bus.

One interesting sight was the crashed remains of a multi stage rocket embedded in the sand. Their tour guide explained that the intrepid explorers from another planet on board this crashed ship perished in the crashed landing, which occurred in the Earth year of 1944.

The guide would not specify which planet the rocket originated from. But, it is not hard to guess it probably was from Germany, which had their own space program during World War II. Other planets would probably not be using obsolete and inefficient rocket technology.

After a day spent on the Moon, the Earth contactees were returned to their homes on Earth.

Menger was also shown how to create a radio controlled flying saucer by his ET friends, which he called an Electrocraft, as described in his book, *The High Bridge incident: The story behind the story*. Eventually, Menger was asked by the FBI to not build any more Electrocraft for reasons of National Security. He, being a patriotic citizen, agreed not to.

For those of you exclaiming it is too hot to live on Venus, retired Lieutenant Colonel Wendelle Stevens has an explanation. He worked at Air Force Intelligence and became a UFO researcher after retiring. His explanation is that the Venusians live on the Astral plane and must "densify" and lower their vibratory rate to visit Earth.

While this may be true, I have the feeling that NASA has not been totally truthful about conditions on Venus or on other planets, like Mars for example.

Another interesting case which Wendelle Stevens investigated of another Venusian visiting Earth and living among us, was that of Omnec Onec.

Her life is described in *Omnec Onec: Ambassador from Venus* by Omnec Onec. She says that Earth is now going through a phase of evolution that Venus experienced millennia ago when they also experienced greed and war and a small class of financial elite who controlled everyone else.

This book is another valuable read to learn of the universal spiritual laws that our ET brothers and sisters wish us to learn, so that we will learn to live in peace and harmony with each other and not destroy our planet through senseless and unnecessary war.

Omnec Onec has a website here: http://omnec-onec.com/

One universal law is that more advanced civilizations are not to interfere in the evolution of lesser evolved civilizations like that of Earth. One exception is they may interfere to prevent a lesser evolved civilization from destroying their planet. This is now threatening Earth - with unenlightened world leaders having nuclear weapons at their disposal. Our ET visitors have disabled Nuclear ICBMS on many occasions as a number of military personnel, manning the ICBM silos, have testified.

Also the famous contactee, George Adamski stated that he was taken aboard Venusian saucers. So, apparently Venus has an advanced civilization that exists on higher planes of existence, if not on the physical, if these contactees and ambassadors from Venus are to be believed.

A more modern contactee is Steven Greer, MD. His first encounter with a flying saucer was as a nine-year-old child, which he saw while playing with his friends. For nights afterwards, he would have dreams of communicating with the ETs and considered the whole affair as not that unusual.

At age 17, he had a near death experience. While his conscious was out of body, he experienced infinite awareness, joy and the perception of an endless perfect creation. Then, he was approached by two avatars appearing as two brilliant points of light. His consciousness became one with the avatars and non-verbal *knowledge* was transferred to him.

Eventually, the communication became more linear and he was verbally given the choice of going with them or returning to Earth. He asked them which they preferred. They responded that they preferred that he return to earth and do more. But, the choice was his. The experience was so beautiful that he really did not wish to return to Earth. But he intuitively realized that he should accept Divine will and agreed to return to Earth.

On fully returning to his body, he realized that he was cured of the sickness that had caused this experience. Thereafter he understood that death is not to be feared because it is merely a transformation from one state to another and our conscious is immortal.

At age 18, he took up Transcendental Meditation and quickly was able to return to the Samadhi bliss of his former out of body experience. About a year later, he hiked up to Rich's Mountain, in the Blue Ridge Mountains about 5,000 feet above the town of Boone, North Carolina. There he had another sighting of a flying saucer.

This time, he was taken aboard the craft for a ride. They were out in space and suddenly, the walls of the craft became invisible. He felt like he was suspended in space with nothing around, looking at the glorious universe. Then, the ETs which were only about 3 feet tall, and himself joined in a wonderful group meditation.

In this meditation, he rose above intellect and ego, and understood that everything in the universe was a part of *one* consciousness. There is no "us and them". This is just an illusion of the animal mind and physical body. In the higher state of mind, we are all one! This is a basic secret of telepathy.

After this experience, Steven Greer would use all he had learned from the ETs. He would often go into meditation in bed before sleeping and develop an omnipresent state of conscious. In this state he would reach out to the ETs and visually show them how to reach his location from space. These techniques were later developed into his CE-5 initiative.

CE-5 is close encounters of the fifth kind, where a contactee actually requests a contact with ETs. Steven Greer would, on many occasions actually be able to call in the ET flying saucers using these meditative, higher states of mind techniques.

There would be times when his roommates would excitedly exclaim "Did you see that? There is a space ship right outside our window!" And, Greer would sheepishly reply, "Yeah, I sort of invited them."

He also learned from experience to not go overboard on this ability because many people become quite frightened when actually seeing a UFO or ET being.

Later, Steven Greer would organize the Center for the Study of Extraterrestrial Intelligence or CSETI.

He would train CSETI members in the meditative CE-5 protocols and training to remove fear on actual ET contact. Often a ET craft would be called in, using CE-5, only to have a member become afraid. And, this fear would make the craft retreat, because the ETs didn't wish to cause fear.

Steven Greer personally knew and met with many high level people in government. These contacts were sometimes used to gain inside information of secret projects. Also, after Greer became more well-known, some people working in "black projects" would also disclose what they knew to him.

Eventually Steven Greer organized the Disclosure Project to inform Congress and the people of the U.S. of secret programs that were so secret that often the U.S. President or the Joint Chiefs of Staff at the Pentagon did not know about them.

These Unacknowledged Special Access Projects (USAP) were being carried out by the Defense Contracting Corporations, wthout he official knowledge or approval of the U.S. government, acting as a "break-away government." Their financing came from funds that disappeard through creative bookkeeping from aknowledged defense projects.

This is easier to understand when one considers that these corporations had been taken over by the Nazi International after a seret treaty signed by President Eisenhower.

In addition to the official secret space program run by the Navy known as *Solar Warden*, there were parallel programs carried out in secret by the Defense Contracting companies, as revealed by William Tomkins in some of his interviews. These coporations in effect had created a break-away government. And their goal was to colonize, mine and industrialize outer space for their own profit.

Dr. Greer has interviewed hundreds of government, military and corporate witnesses to UFO events and leads groups like CSETI to make peaceful contact with ETs and is working to reveal technologies that will get our civilization off of fossil fuels. The technology behind how UFOs get to earth hold the secret to how we can sustainably live on earth without pollution and poverty. This technology is one reason for all the secrecy.

The Disclosure Project has over 500 government, military, and intelligence community witnesses testifying to their direct, personal, firsthand experience with UFOs, ETs, ET technology, and the cover-up that keeps this information secret.

For much more knowledge about Steven Greer, CSETI and the Disclosure Project here are the web sites:

http://www.CSETI.org

http://www.DisclosureProject.org

Also Dr. Steven Greer has also authored several books which are on my recommended reading list:

(1) *Extraterrestrial Contact: The Evidence and Implications*

(2) *Disclosure: Military and Government Witnesses Reveal the Greatest Secrets in Modern History,*

(3) *Hidden Truth – Forbidden Knowledge,*

(4) *Contact: Countdown to Transformation.*

The CIA, which controlled MJ-12 and much of the secret space program, went to great lengths in spreading disinformation about the UFO contactees, from the 1950s to the present, from George Adamski to Steven Greer, in order to debunk and discredit them.

This was part of their general policy to debunk the whole UFO phenomena in general, to keep mankind in spiritual darkness and ignorance of universal spiritual law and of our brothers and sisters on other planets and star systems.

This was going on even while the crashed UFOs were being brought to Wright Patterson AFB for analysis and back engineering, and later to Area 51, Sub Area S-4, which was controlled by the CIA.

Technology for the Secret Navy Space Fleet

The German *Vrill* and *Thule* secret societies had originally developed antigravity technology themselves in the 1920s based on channeled information from Aldebaran by medium Maria Orsic. Also, Victor Shauberger was developing his *Repulsine* and *Trout Turbine,* based on his close observations of trout swimming up waterfalls. These machines had both free energy aspects and levitation aspects and were first considered by the Nazi SS for submarine propulsion and then later for flying saucer propulsion. In addition, German spies had recovered Nikola Tesla's 1890s work on an unpatented flying machine he developed that worked on electric – not aerodynamic principles.

The Nazi SS also were given working saucers that could operate in the atmosphere, under water and in outer space by the Draco Reptilians. Also, the *Andromeda* cigar shaped mothership was built that was a space faring aircraft carrier that carried the *Vrill* and *Haunebu* saucers inside.

The *Andromeda* mothership design was obtained by U.S. Naval Intelligence During World War II, and the concept was later used by William Tompkins to design his two-kilometer-long space craft for the Navy at the highly secret Advanced Design section of Douglas Aircraft.

This design process started in 1954, after the Navy put out a sole request to Douglas Aircraft for a proposal for exploratory star mission vehicles. These huge craft were deemed necessary as a high priority project to defend planet Earth from hostile extraterrestrials. Also, they were to be used to explore the Solar System and beyond.

The propulsion for this ship would be electromagnetic, using concepts developed by Dr. Wolfgang Benjamin Klemperer. Klemperer was born in Dresden Germany on January 18, 1893 and in 1924 emigrated to the U.S., where he became employed by Goodyear-Zeppelin Corporation in Akron, Ohio. From 1936 onwards, Klemperer worked for Douglas Aircraft Company, Santa Monica, California.

A concept that was theoretically developed by Dr. Klemperer and independently by Dr. Burkard Heim allowed for unique propulsion possibilities. Dr. Klemperer's theory was closely identical to that of Dr. Heim's. Dr. Burkard Heim is a genius, considered on a par with Steven Hawkins in Germany. Little information can be found about Dr. Klemperer's theory. So, Dr. Heim's theory is discussed here.

Heim started with Einstein's General Relativity Theory, but modified it for application in the microscopic range. Here, the field equations become eigenvalue equations. For invariance reasons Heim had to introduce a 6-dimensional manifold. The existence of a smallest area required the computation with differences rather than with differentials, and with selectors instead of tensors. According to Heim, Einstein's assumption of one single metric was too simple.

He introduced three partial structures, which constitute four possible metrical tensors by correlations.

This complicated geometry leads to 1,956 eigenvalue equations from which it is possible to deduce the mass spectrum of elementary particles and to describe their internal structure fluxes. Matter consists of an exchange of maxima and minima of condensations of the smallest areas in subspaces of an R6 space.

Contrary to vacuum fluctuations, matter exists when the geometrical exchange processes always return to their starting point. These geometrical fluxes produce a spin. Since this spin tends to stay orthogonal to the vector of world velocity, each acceleration leads to a resistance force or inertia.

If the meso-field equations are approximated so that the 4 dimensional space time continuum results, and the matter fields are restricted to their electromagnetic characteristics, two states of this meso-field result which Heim designated the dynabaric state and the contrabaric state.

The dynabaric state can act on the gravitational field to directly convert it into electromagnetic energy. The Contrabaric state operates on the electromagnetic field to directly produce gravity or acceleration of material objects.

There are several possible ways to generate gravitational fields and gravitational waves in Heim's theory. A theoretical possibility consists in the generation of gravitons from neutrons. The generation of acceleration fields was later investigated by the German spaceflight company DASA. Heim himself proposed to test the contrabaric effect predicted by his theory.

This 6 dimensional field contains not only electromagnetic and gravitational phenomena but also a third phenomenon called a meso-field. This meso-field is capable of 2 different effects.

The contrabaric state transforms a material phenomenon directly into propulsive action by emission of gravitational waves, resulting in accelerated motion.

It also transforms material phenomena into the dynabaric state through which purely electromagnetic energy is liberated from matter without any waste product or heat.

Propulsion is achieved by first liberating electromagnetic energy and then transforming it directly into accelerated motion.

Another one of Dr. Klemperer's electromagnetic design concepts utilized the concept that once a transverse electromagnetic wave is emitted from an antenna, its electric and magnetic field effects become independent of the antenna and are then embedded into the ether of space itself, while travelling at the speed of light away from the antenna.

An antenna of suitable geometry in the front of the ship emits EM waves backward and insulated conducting plates or surfaces along the ship are wired to the proper electronic switching circuits connected to charge storage capacitors. If these plates have their charges varied from positive to negative in proper synchronization with the EM wave, traveling in the space just outside the ship, propulsive forces can be generated. This can be understood by visualizing that positive charged plates are repelled by an electric field oriented as one coming from a positively charged body and attracted to the opposite direction of electric field. A similar effect applies to negative charged plates.

This was by no means the only electromagnetic propulsion

method considered by the scientists and engineers at Advanced Design, where the electro gravity concepts developed by Thomas Townsend Brown were also well known. Brown experimentally discovered that a sufficiently high electric field warps space-time and creates artificial gravity which can be shaped to propel space ships. This electric space-time warping is vastly more efficient than that caused by mass alone.

Then, there was also the warping of space time caused by rotating magnetic fields as utilized in the Philadelphia experiment (or Project Rainbow as it was officially named). These rotating magnetic fields could create traversable wormholes and cause teleportation.

In any case, the Advanced Design people worked from a Manual titled: THE INCREDIBLE DOUGLASS MTM-622 UNCONVENTIONAL PROPULSION SCHEMES. The MTM-622 was highly classified then and now. It was the bible of the secret spaceship designers at the Douglas Advanced Design section in the 1950s and only recently are some of the concepts therein being discussed at DARPA.

In the 1960s, Lockheed Aircraft, at their Skunkworks, in a highly classified project, was developing the Alien Reproduction Vehicle (ARV), also called the Flux Liner, Information on the ARV accidently leaked out in November of 1988 at an air show held at Norton AFB.

Brad Sorenson attended the air show and inadvertently found himself in an area away from the crowds that was intended only for the top military brass. In a hangar, a curtain was pulled away, revealing three different sized antigravity craft, all hovering in the air!

They looked like truncated cones with the large part down and with a hemisphere on top. They had respective bottom diameters of about 24 feet, 60 feet, and 125 feet. In front of the exhibit was a cut away model of the smallest craft with notes explaining the functions of the interior portions. Brad Sorenson also overheard the military officer explaining that the ARV's capabilities included faster than light travel.

The crew compartment of the smaller 24-foot craft was about 12 feet in diameter and was equipped with seating (ejection seats) for four occupants, mounted "back-to-back" in a circular arrangement around a central column. Among the notes, it was stated that The Flux Liner accessed "zero point" energy which caused mass cancellation of the structure of the craft and that it was capable of faster than light speed. All three craft looked "well used" like they had been in service for quite a while.

Brad Sorenson was an old college friend of Mark McCandlish, a well-known, aerospace industry illustrator. In fact, it was McCandlish that informed Brad Sorenson of the Norton AFB air show in the first place. They were planning to go together but McCandlish had an important assignment come up and could not attend. So, Sorenson went alone.

McCandlish kept perfecting drawings of the ARV from detailed information passed on to him by Brad Sorenson. Still, keeping a healthy skepticism, McCandlish did some of his own investigation into the matter. He finally was able to verify Sorenson's story from his congressman, Representative George E. Brown, Jr., who was the Chairman of the Committee on Science, Space and Technology at that time. A staff member in the congressman's office confirmed that the event did take place and that there were, in fact, three discs at the exhibit.

Then, McCandlish was a total believer and started making very detailed pictures of the craft and researching further. At the 2001 *Disclosure Project* National Press Club Event, McCandlish showed photos of an ARV taken in 1966 in Provo, Utah. It appeared highly identical to his drawings. There, he also explained the story behind his illustration of a type of man-made faster than light flying disc craft called an "Alien Reproduction Vehicle" (ARV).

After the *Disclosure Project*, the ARV became widely known about. Here is a documentary: https://www.youtube.com/watch?v=ht3zkDxAjUM

And here is a video on the subject: https://www.youtube.com/watch?v=GEfS25_zyHE

So, there seemed to be a variety of spaceship propulsion techniques available to the covert aerospace industries to choose from.

Tompkins says that he approached his work by studying the mission parameters for the requested future space battle groups. He then was able to come up with designs that would allow the Navy to fulfill its future space missions.

Creating the configuration of a Naval Space Battle Group comprising kilometer-long vehicles from the mission parameters he had been given, Tompkins explains:

> "I redefined a standard Naval space battle group complement, stating that it would consist of one 2.5 kilometer spacecraft carrier, with a two-star on board as flag, three to four 1.4 kilometer heavy space cruisers, four to five 1 kilometer space destroyers, two 2 kilometer space

landing assault ships for drop missions, two 2 kilometer space logistic support ships, and two 2 kilometer space personal transports."

Eventually, there were eight of these space carrier battle groups that were built for the U.S. Navy in the 1980's and 1990's, according to Tompkins.

Although the designing and planning started in 1954 at the highly secret Advanced Design section at Douglas Aircraft for the Navy, many concept models and testing processes were required to finally get a production model created. After his initial designs of the space carriers were completed in the early 1960s, Tompkins claims that it took nearly a decade for detailed architectural plans to be developed, enabling official construction to begin.

Also problematic, was to locate a secret production center that could construct Naval space craft that were kilometers long. One kilometer is 3,280.84 feet. The largest known Navy ship is the super carrier, USS Nimitz, measuring 1,092, feet long and weighing 101,600 tons which was built at the Newport News shipbuilding yard. So, some of these Naval space ships would be over 6 times longer than the Nimitz!

A very large natural cave was discovered in the Wasatch mountains North East of Salt Lake City, Utah. Part of this cave was converted into a naval space ship fabrication factory large enough to build a 2.5 kilometer sized craft. This enormous underground facility has a door which is the size of some 40 city blocks which opens from the surface into the immense underground facility below.

The cave construction site could have been expanded to

build 25 kilometer ships if funding was available according to Tompkins. Consequently, building began in the 1970's and the first operational space carriers were deployed in the late 1970's, under a highly classified space program called Solar Warden.

The work was done cooperatively on the giant space crafts by thousands of employees from Boeing, Lockheed, Northrop, and Rockwell corporations.

This noncompetitive cooperation between these different companies was a feature of the fascist efficiency of these companies after they were taken over by the paperclip Nazis in the 1960s and told that the defense of Earth depended on their cooperative working together.

According to William Tomkins, the door mechanism used to launch the gigantic space craft is so precise and seamless that when it opens and closes, anything on the desert floor, including the plants and animal life which may be on that door, is not disturbed nor harmed in the slightest. With the door closed, any hiker in this very remote location could walk right over it and not even know that it was there.

There were parallel programs also going on where smaller space craft were also being built. Newport News shipping yard was used to modify some nuclear submarines being built there. The nuclear reactors were taken out and replaced with antigravity drives. A lot of other equipment was also replaced. These highly modified subs were launched and while at sea could be secretly launched into space.

Also, much smaller saucer like antigravity craft were being built by Northrup Grumman at the "Ant Hill" underground facility in the Tehachapi Mountains near Lancaster, California.

Furthermore, there also was a highly secret cooperation between the U.S. and USSR all during the Cold War in a joint program to defend the Earth from hostile ET invaders. Many of the weapons programs on each side of the arms race had funds siphoned off to finance these secret space programs.

So, the seeming hostility between Russia and the West was advantageously used to convince the citizens of both sides that the astronomical weapons spending was necessary for their survival, while a good percentage of the money to fund these weapons was, via creative bookkeeping, being used to fund these secret space programs that these citizens had no knowledge of.

World leaders realize the advantage of having an "enemy" to unify the public behind the government that this public think they need to protect them.

To the person accustomed to the reality picture provided us by the corporate controlled news media, this may all sound like science fiction. But, to researchers of UFO incidents and secret government agencies, like the Nazi SS, CIA and MJ-12, this picture becomes quite plausible when all the pieces of the puzzle are fitted together.

I have been one of those researching this sort of material since 1957 when I read books about Adamanski's flying saucer encounters, and newspaper accounts of Otis Carr's flying saucer, experiments in Hesperia, California. Carr claimed his saucer could reach the Moon in 12 hours,

Deep Underground Military Bases (DUMB) started with the Germans during World War II, as the Allies were bombing the hell out of everything on the surface. Entire factories were moved underground, including the V-2 assembly line at the

Mittelwerk in Nordhausen.

After the war, the U.S., Russia, Canada, and Britian got much of the German technology. The German patent office was raided and literally tons of paperwork with the patents were confiscated and transferred to Wright Patterson Army Air Force Base AMC (Air Material Command) Alien Technology Division, for analysis. This was in addition to all the other sources of intelligence on German technology.

Members of the German engineers that built the Mittelwerk were sent to Wright Patterson to design underground facilities there that house the recovered German and extraterrestrial technology. These men included Xaver Dorsch who replaced Fritz Todt of Germany's Todt Organization which designed many of Germany's underground facilities.

From these underground bunkers, this technology was back engineered, largely by Operation Paperclip Nazi engineers and scientists brought over from Germany. These included men like Siegfried Knemeyer, the former head of the German RLM (The Reichsluftfahrtministerium), the Third Reich's Air Ministry for aircraft development for the Luftwaffe, and Dr. Hans Amtmann, an expert in vertical takeoff aircraft, and Dr. Alexander Lippisch who was well known and a pioneer in tailless aircraft, the US Delta wing fighter, the F-102A Delta Dagger and an advanced design of a ground effect flying boat.

These men also did a reverse engineering on the object that crashed near Roswell, New Mexico as well as recovered German saucers. It was not that our scientists and engineers were incapable. It was because the German technology was based on "other science" developed in Germany, like Victor Shauberger's *Repulsine* levitation science, Karl Schappeller's glowing magnetism and primary state of matter, or torsion field mechanics which could create levitation and inertia shielding around a craft, which was incomprehensible to our scientists trained in conventional science.

When back engineering some of the extraterrestrial craft, the engineers and scientists discovered a particularly amazing thing. The material of the craft seemed alive!

Not only was the craft able to self-repair its damage, like a wounded animal could heal its wounds, although faster, but also the craft could respond to and be controlled by the thoughts of its pilot. This amazing extraterrestrial technology seemed to be based on a combination of nanotechnology and biology. It definitely was based on a manufacturing process that amplified the inherent consciousness imbued into matter by ether. These discoveries quite likely led to the science fiction TV series, *Farscape*.

When the Vrill medium, Maria Orsic was giving Dr. Otto Schumann her channeled information to build the JFM series of saucers, at one point she delivered plans for a mental controller headband so that these saucers could be controlled by the thoughts of the pilot. Later, the *Vrill 7 Giest* saucer could be flown telepathically by Maria Orsic - without the headband! So, even the Germans were able to develop some of this technology which they telepathically received from beings on Alderberan in the 1920s.

Also, most of the recovered documents after the war, were in German and had to be translated by German speaking people. Many of the words had no equivalent in the English language and a new technical dictionary had to be created at Wright Patterson so U.S. engineers and scientists could understand what they had captured.

According to Colonel Philip J. Corso, in his book, *The Day After Roswell*, in one area of Wright Patterson, parts from these craft were scattered on the floor and experts from different industries with the proper security clearances were brought in to examine the parts and back engineer them.

They were told that they could have the patent rights to any technology copied or developed from these parts. For example, Bell labs got the transistor patents from back engineering parts of extraterrestrial crafts according to Corso. This policy would cause greater incentive to the industry and, at the same time, hide the actual source of the technology. Of course, most technology of special interest to the military would become highly classified and not revealed to the public

Corso was a Special Assistant to Lt. General Arthur Trudeau, who headed Army Research and Development, and was in charge of the Foreign Technology Desk. Corso's book, *The Day After Roswell*, is another expose of the official lie about the Roswell incident.

Because of the Post War threat of nuclear attack, especially after Russia demonstrated their own nuclear capability, a program of constructing deep underground military bases or DUMBs accelerated. The NORAD center built into Cheyenne Mountain, Colorado, was designed to be hit with a Hydrogen Bomb and still keep functioning.

William Tompkins claims he designed the underground base at Offutt AFB, which is home to the 55 Wing. Its mission is to provide dominant intelligence, surveillance, reconnaissance; electronic attack; command and control; and precision awareness to national leadership and warfighters across the spectrum of conflict any time, any place.

The underground portion of the base is 300 feet underground and has a 28.5-foot-thick reinforced concrete dome over it, according to Tompkins, to also make it Hydrogen bomb resistant.

These underground bases are known about. But there are more than a hundred DUMBs in the U.S. that are kept quite secret from the public, as whistleblowers like Phil Schneider and researchers like Richard Sauder with his books, like *Underground Bases and Tunnels: What is the Government Trying to Hide?* have shown.

A man well connected to military black projects, John Lear, had this to say in 2003 in an interview on the Art Bell *Coast to Coast* radio show:

> "There are tunnels and base complexes that connect China Lake, California City, Norton AFB, Tonopah, Groom Lake, Nevada Test Site, Los Alamos, Dulce, NORAD, Oklahoma and to the East."

The method of constructing these underground tunnels is described in an article from Veterans Today on February 10, 2013 titled *Elite Underground*:

> "Imagine a machine tunneling seven miles per day through solid rock, boulders and clay...virtually anything below the waterline. A machine whose heart is a compact nuclear reactor circulating liquid lithium at 2,000 degrees F. through a rotating face that melts a tunnel 40 feet in diameter; even injecting magma into fractures in bedrock for extreme solidity, sealing the tunnel

with a glassy lining, and leaving no excavated material behind. Imagine a subterrene. The nuclear subterrene was born at Los Alamos National Laboratory in New Mexico. Patents were taken out in the 70s.

And since their mission is a menace to public interest, subterrenes have been minimized from public awareness. It might be safe to say that subterrenes are the hottest tool going, underground. A kind of nuclear nightcrawler."

Phil Schneider, who had an extremely high security classification and worked on these DUMBs, stated that there were 129 DUMBs in the U.S. in 1996. But, there probably have been more constructed since then. They were on the average over a mile-deep underground and their size was between 2.66 and 4.25 cubic miles in size and they are connected by tube shuttles. Tunnel boring machines that can drill about 7 miles per day through solid rock are used to make these tunnels.

Also, Schneider revealed an incident where he was drilling holes to create an underground base under the Archuleta mesa near Dulce, New Mexico. They apparently drilled into an already existing extraterrestrial underground base. What is called the "Dulce wars" followed, where many U.S. special forces were killed. According to Phil, only himself and 2 others survived this war with tall Grey extraterrestrials. (7)

Phil Schneider also revealed that he had viewed ETs at Area 51 and that he had attended the real UN meetings, not in New York, but in deep underground military bases or DUMBS. After his second meeting there, realized that ETs were actually running U.N. policies and he decided that he was working for the wrong people. This is what led him to go public with what he thought was really going on based on his own experiences.

In his last reordered lecture, Schneider told his audience there had been 13 murder attempts on his life by government agents intent on preventing him continuing to inform the public of the existence of ETs. He said that he was speaking out because, "I love my country more than I love my own life."

Phil Schneider had been on a speaking tour with his revelations in 1996, when suddenly he had supposedly committed suicide. I say "supposedly" because it was a murder certain authorities didn't wish to be investigated. They usually claim a suicide in these cases because suicides are not thoroughly investigated - if at all - the Vince Foster "suicide" being a classic example.

So, after knowing all this, we see that the huge underground base, in the remote wilderness of the Wasatch mountains North East of Salt Lake City, where two-kilometer-long Navy space ships were secretly built does not seem so incredible after all.

This large underground production facility in the Wasatch Mountains was managed like other "black projects". Nazi efficiency was used in maintaining secrecy, using the compartmentalized sealing of information from people on other parts of the program on a "need to know" basis.

Employees signed draconian secrecy agreements that could be punishable by, not only by loss of employment, but 10 years in prison for violating National Security, if breached. The few whistleblowers that spoke of these things usually had all their records disappear, ruining their careers, or they just disappeared themselves.

Many of these employees lived in Southern California. The

Palmdale area replaced Santa Monica as a hub of secret "black projects technology" in the 1980s. The Lockheed Martin "Skunk Works" was transferred to site 10 at U.S. Air Force Plant 42 in Palmdale in 1989. Northrop Grumman created an underground facility known as the "Ant Hill" in the Tehachapi Mountains North West of Lancaster. And secret antigravity testing was occurring nearby at Edwards AFB. China Lake Naval Air Weapons Station also had a large underground facility where secret Navy projects were going on.

So, there was a large cadre of "black project" employees with high security clearances in that area, and some were shuttled in these underground tube systems at supersonic speed between underground bases near home in Southern California and work in the large Naval space ship construction underground facility in Utah.

The funding for this gigantic, secret, underground spaceship construction facility was largely obtained from creative bookkeeping with the Pentagon budget. To see how this works, we have present day revelations by a Reuters investigation on missing Pentagon trillions of dollars:

The Defense Finance and Accounting Service (DFAS), the behemoth Indianapolis-based agency that provides finance and accounting services for the Pentagon's civilian and military members, could not provide adequate documentation for $6.5 trillion worth of FY 2015 year-end adjustments to Army general fund transactions and data.

The DFAS has the sole responsibility for paying all DOD military and personnel, retirees and annuitants, along with Pentagon contractors and vendors. The agency is also in charge of electronic government initiatives, including within the Executive

Office of the President, the Department of Energy and the Departing of Veterans Affairs.

While there is nothing in the IG's report specifying that the money has been stolen, the mere fact that the Pentagon can't account for how it spent the money reveals a potentially far greater problem than simple theft alone.

For every transaction, a so-called "journal voucher" that provides serial numbers, transaction dates and the amount of the expenditure is supposed to be produced. The report specifies that the agency has done such a poor job in providing documentation of their transactions, that there is no way to actually know how $6.5 trillion dollars has been spent. Essentially, the government has no way of knowing how the Pentagon has spent the trillions of taxpayer dollars allocated by Congress for national defense.

In turn, employees of the DFAS were routinely told by superiors to take "unsubstantiated change actions" commonly referred to as "plugging" the numbers. These "plugs" – which amounted to falsifying financial records – were then used to create the appearance that the military's financial data matched that of the U.S. Treasury Department's numbers when discrepancies in the financial data couldn't be accounted for, according to the Reuters investigation.

According to that Reuters investigation:

For two decades, the U.S. military has been unable to submit to an audit, flouting federal law and concealing waste and fraud totaling billions of dollars.

Linda Woodford spent the last 15 years of her career inserting phony numbers in the U.S. Department of Defense's accounts.

Every month until she retired in 2011, she says, the day came when the Navy would start dumping numbers on the Cleveland, Ohio DFAS.... Using the data they received, Woodford and her fellow accountants there set about preparing monthly reports to square the Navy's books with the U.S. Treasury's.... And every month, they encountered the same problem. Numbers were missing. Numbers were clearly wrong. Numbers came with no explanation of how the money had been spent or which congressional appropriation it came from.

While many of the problems occurred due to bookkeeping errors rather than actual financial losses, the DFAS has failed to provide the necessary tracking information essential to performing an accurate audit of Pentagon spending and obligations, according to the IG's report.

"Army and Defense Finance and Accounting Service Indianapolis personnel did not adequately support $2.8 trillion in third quarter adjustments and $6.5 trillion in year-end adjustments made to Army General Fund data during FY 2015 financial statement compilation," wrote Lorin T. Venable, the assistant inspector general for financial management and reporting. "We conducted this audit in accordance with generally accepted government auditing standards."

The Pentagon has a chronic failure to keep track of its money – how much it has, how much it pays out and how much is wasted or stolen. Adding to the appearance of impropriety is the fact that thousands of documents that should be on file have been removed and disappeared without any reasonable explanation.

DFAS "did not document or support why the Defense

Departmental Reporting System . . . removed at least 16,513 of 1.3 million records during Q3 FY 2015. As a result, the data used to prepare the FY 2015 AGF third quarter and year-end financial statements were unreliable and lacked an adequate audit trail," according to the IG's report.

On top of these accounting problems, we had the statement by Donald Rumsfeld on 9/10/2001 that the pentagon couldn't account for $2.3 Trillion and then the next day the Pentagon was hit in the Naval Intelligence section that was investigating financial crime, destroying all their records and killing most of their investigation team.

The Solar Warden program of the 1980s got largely funded by this type of "creative accounting" with President Reagan's Strategic Defense Initiative or "Star Wars" budget. The U.S. national debt tripled during the Reagan Administration. But, maybe it was worth it to get all this secret exotic technology created.

Also, illegal narcotics trafficking was used to finance these "black projects", that were so secret that very few in our constitutional government knew anything about them. Much of the following is documented in *CIA: Crime Incorporated of America* and other well researched books.

The desire to maintain a presence in Vietnam was actually to control the flow of heroin and illegal drugs into the world market. The main players in the orchestration of this program where George "Poppy" Bush Senior, a.k.a. George Scherf Jr., Nazi spy/ Nazi New World Order administrator. There are so many books and evidence to this it is ridiculous. Chip Tatum, Brice Taylor and Gen. Westmoreland, Kissinger, Robert Gates, Richard Armitage, George Schulz, Cheney, Rumsfeld, and many others, were and are deeply involved in this legal illicit drug trade.

The money was laundered through the Ameri-Asian and Nugan Hand Bank and later, through BCCI. This same model was used in the cocaine trafficking routes in South America with many of the same players, especially the Bush family, at the helm. The Clintons were heavily involved as well, as a major supply route was established through Mena, Arkansas. The Clintons financed their political careers through this money and that was laundered through the Arkansas state agriculture agency and was almost exposed in the political Whitewater trial. That trial was tainted by the murder of Vince Foster, who possessed much incriminating evidence of the Clintons, in the White House! The Clintons are complicit in crimes against humanity and will hopefully be soon be brought to justice.

The horror and the suffering and the untold wasted and destroyed lives, not only from the war itself but from the millions of drug addicts and destroyed families, weighs heavily upon the karma of United States and the elitist secret government. These military leaders in charge became very wealthy and amassed tremendous power through their actions, devotion and loyalty to this secret funding program.

At the highest levels this was reasoned out and justified by military minds that we needed money to develop a secret space program to protect us from aliens we could not control and who had higher technology. This was compartmentalized under the NSA (National Security Agency) that trumps the CIA and even the U.S. President. Private individuals and corporations control the NSA and other world government security agencies, media and the entire world's financial structure.

Above these oligarchs are galactic criminals of extraterrestrial origin, like the Draco Reptilians, who have secretly gained control of the world's financial, societal and political structures.

So through a secret organization, the United States government was taken over by vested corporate interests which allegedly were working with an altruistic motive to protect the earth from what they considered hostile an alien invasion. This is how it was sold to the military anyway. The generals and security types who had the clearance to know about the ET presence had to choose the lesser of two evils. This erroneous policy and thinking has gotten us up to our necks in lies and deceit and must now come to an end.

Another technology used in the Secret Space Program is Teleportation and Time Travel. Nikola Tesla was the first to experiment with this type of technology after receiving an accidental high voltage electric shock. For the brief time, he was paralyzed by this shock, he seemed to be stuck in a strange space where the past, present and future were all happening simultaneously! His lab assistant saw what was happening and quickly threw the power switch turning off the high voltage and saving Tesla's life.

After that close call, Tesla carefully experimentally performed more tests to discover what had actually happened. He discovered that rotating magnetic fields could not only induce electric current but also could mysteriously affect space-time.

In later years, Tesla, Thomas Townsend Brown, Albert Einstein and John von Neumann, among others, were hired by the Navy in *Project Rainbow*. Tesla believed the crew could would be harmed by the experiment and wanted more time to work out the "bugs". The Navy said that it was a wartime emergency and they were going to continue with a "live" experiment. Tesla then resigned from the project.

Nevertheless, they were able to place a destroyer escort, the

Eldridge, into hyperspace and cause it to disappear from normal space time. Eventually the *Eldridge* reappeared. But the disastrous effects on the surviving crew members caused the Navy to shut the project down.

However, there were other parties that realized that they were on to some important science and technology and decided to continue experimenting in a program named *Project Phoenix* with John von Neumann as project leader. This would lead to a number of secret experiments at Brookhaven National Laboratories and other locations.

Also experiments were carried on at the abandoned Fort Hero, Long Island called the *Montauk Project*, by some in the early 1980s. This project actually was the third iteration of *Project Phoenix* called *Phoenix III* by its directors. This project experimented with mind control, physic energy, time travel and teleportation using microwave signals from a Sage Radar System. (8), (9)

A device referred to as a *Temporal Vector Generator,* or TVG was developed. The TVG was a device to *navigate* 3D time, just like you would target an object in space except they target a temporal coordinate. Once you have the coordinates of where you are and where you want to go, you can plot a course from one to the other— using a traversable wormhole tunnel. The TVG, itself, is not much more than a surveyor's instrument for the temporal landscape.

The cosmic sector of reciprocal time space was an area where psychic phenomena occurred. A clairvoyant could see what would happen on the future timeline in 3D time, just as a normal person could see a car coming down the road in the material sector. Two persons at the same time coordinate could communicate telepathically even if separated a great spacial distance. Psychics have developed their sensory perception to a greater of lessor degree in the cosmic sphere or 3D time sector of existence which is the area where consciousness and life itself is formed.

This could be why women, the bringers of life into this world, usually have higher psychic ability or intuition than men who have to struggle in the material sector of existence.

In any case, psychics were highly valued in these secret projects dealing with time travel and teleportation.

Those of you familiar with the book, *Montauk Project: Experiments in time* by Peter Moon and Preston Nichols, know about the Montauk Chair which the psychic, Duncan Cameron would sit upon while psychically driving the Sage Radar system via Helmholtz coils arraigned around the chair on all three x,y and z axis. These coils would "read" his mind and then feed signals into a computer system controlling the output to the Radar.

One worker at Ft. Hero on the *Phoenix III/ Montauk Project*, from 1981 to 1983, was a man who was hired because of his high technical and psychic abilities and prefers for reasons of personal safety, to just be named Daniel, relates the following about the Montauk Chair:

> "The Orion Cube was the core of the Montauk Chair, used to direct temporal experiments. The cube is actually the navigation computer from a SM [Saurian Man or Draco] spaceship, probably one of the smaller scout vessels. Consider that in order for a spaceship to travel faster-than-light (FTL) without time dilation, it must be able to successfully navigate through coordinate space

and coordinate time, simultaneously. Essentially, the "cube" provides a window into both realms for the pilot to navigate with.

What happens with FTL travel is that as you pass the speed of light (the EM barrier), the spatial dimension in the direction of travel shifts into the coordinate time realm, as speed is a reciprocal relation between space and time—cross the barrier, and you start having a temporal dimension instead of a spatial one. But, since time is 3D, not a 1D vector, you have to keep the ship going straight in the coordinate time realm, which means you need to be able to see the realm to properly navigate it.

This is what the Orion Cube does. Upon transition to FTL velocity, the volume of the ship has been altered to 2 spatial dimensions and 1 temporal one—which means that in space, it appears as an area, not a volume, usually a flat disc or saucer shape, depending on the FTL speed. It gets flatter the faster it travels beyond the speed of light (the reciprocal relation: more time, less space). In time, it is cylindrical or jet-like, like a meteor streaking through the temporal landscape that needs an accurate flight path so it arrives at both the correct location in coordinate space, and in coordinate time.

In the Phoenix III project, they still had all the bits of the ship's navigational system, the cube, interface and pilot's seat—the Montauk chair setup described by Preston Nichols, Al Bielek and others. That's what we ended up hooking to the IBM mainframes, which were essentially a simulation of the responses of the spacecraft (though they didn't tell us that). They knew the ships traveled FTL, so they knew the device could somehow affect time and the project was built upon that principle. They were able to fool the navigation system into thinking the craft was accelerating past the speed of light, so it would open the temporal navigation window—the vortex—even though it remained fixed in space. Of course, that created a few other problems as occasionally happens with simulations versus reality, like the odd wall disappearing and strange behaviors of animals in the vicinity.

It appears they either had other cubes from other craft, or pulled the one from Montauk prior to the abandonment of the project in 1983.

I'd bet they just replaced the room full of IBM mainframes with a PC these days—probably a lot more computing power than we had back then—which made the device much more portable. From what Burisch related, it doesn't look like things changed much—they're probably still using my original "chair" driver code I wrote for the IBM! I don't know what they did for the interface, as the chair required a person with psionic ability to use it. I'd guess they reverse- engineered some sensors to pick up the general psycho-emotional activity in the immediate area, to control the projection into coordinate time (without some kind of "pilot," it would be random and useless)."

Daniel also explained that a series of 22 sine wave graphs would show the activity of the person sitting in the Montauk Chair. When that person would still their mind, the graphs would remain still. Then, the technicians would turn knobs and tune the chair to the person by flat lining the 22 sine waves. Then, when the chair was tuned to the operator, the magic would happen!

The operator sitting in the chair could visualize an object like a cold drink of beer and the object or a cold bottle of beer would materialize out of the ether of space! The object would remain solid for a period of time before eventually fading away.

Even more strangely, the operator could think about a particular place in space and time and a portal would open up that would allow travel to that place in space and time!

People, usually homeless children picked up in New York, would be sent through these portals, in a series of trial and error experiments. Many lives were lost before they figured out how to insure their safe return. Portals would also randomly appear at different locations on the Fort Hero base. People eating in the cafeteria would see a wall starting to disappear and would pick up their lunch trays and move to another location in the cafeteria.

Daniel explained much more in a number of lengthy interviews with David Wilcox which is revealed in his recently published book (September 2016), *The Ascension Mysteries*. So, I only cover the main points here.

So, through the various *Phoenix Projects*, the bugs were eventually worked out of the space-time travel technology. Andrew Basiago was one of a number of people who were teleported through space and time in the late 1960s and early 1970s in *Project Pegasus*.

He claims that he was sent back in time and space to President Lincoln Gettysburg Address and later to the Ford Theater when Lincoln was assassinated, in 1972 via a plasma confinement chamber located in East Hanover, N.J. and explains the technology:

"The machine consisted of two gray elliptical booms about eight feet tall, separated by about 10 feet, between which a shimmering curtain of what Tesla called 'radiant energy' was broadcast… Radiant energy is a form of energy that Tesla discovered that is latent and pervasive in the universe and has among its properties the capacity to bend time-space."

Basiago said project participants would jump through this field of radiant energy into a vortal tunnel and "when the tunnel closed, we found ourselves at our destination."

"One felt either as if one was moving at a great rate of speed or moving not at all, as the universe was wrapped around one's location," Basiago said.

And how did these alleged time travelers return to the present day or their point of origin? According to Basiago, some sort of holographic technology allowed them to travel both physically and virtually. "If we were in the hologram for 15 minutes or fewer," he explained, "the hologram would collapse, and after about 60 seconds of standing in a field of super-charged particles ... we would find ourselves back on the stage - in the present"

Project Pegasus ended up using young people (pre-teen) as chronauts because adults psychologically could not handle time travel or teleportation very well. Basiago and other young people were also teleported between New Jersey and Santa Fe, New Mexico. Their hope was to train a cadre of time travelers and teleporters that would become accustomed to it and use them later in life.

This was to happen to Andrew Basiago and others in the 1980s when he was teleported to a Mars base using a different technology given to the military by the Grey extraterrestrials. This machine was called the "Mars Jump Room" by those involved in the project but, officially it was called an Aeronautical Repositioning Chamber or ARC.

One Jump Room or ARC was located in El Segundo near the Los Angeles Airport on Sepulveda Boulevard in the old Howard Hughes building. Another was located in New York at the CIA complex near the George Washington Bridge.

Others included in this Mars Jump room program were Barak Obama (under the name Barry Soetoro), William Stilling, Cortney M. Hunt (career CIA), Regina E. Dugan (who later Obama, as President, appointed head of DARPA) and Bernard Mendez. Now (Oct., 2016), Andrew Basiago was also write in candidate for U.S. president.

William White Crow, US Army (ret.) was also involved with the mars Jump Room program. Here is a statement made by him on Facebook, 4/29/16:

> "I taught Andrew D. Basiago martial arts in 1971-72 when he was time traveling for DARPA. Ten years later, I guarded his life on Mars. I can say from personal experience that Andy's time travel and Mars visitation accounts are true. His life story symbolizes the secret exploits of the US chrononauts. Like thousands of others who have served our great nation secretly, Andy's service to our nation has been profound. Today, it is my honor to endorse Andy for President."

There were others in this program but their names could not be remembered by those coming forward with the information.

In this CIA program, these young people were trained at the College of the Siskiyous near Mt. Shasta, California in a class taught by Major Edward Dames. Obama and Dames both deny being involved, but Basiago, Stillings, and Mendez all say that they were both involved.

So a number of upstanding citizens have come forward and testified that this Jump Room technology is real and well known people were involved. I have stated that President Barack Obama is a deep cover CIA agent, as were his mother, father, stepfather and grandmother in *CIA: Crime Incorporated of America*. The Mars Jump Room Program in the 1980s was being run by the CIA as well as MJ-12 and the security for Area 51.

Possibly the memories of Obama and Dames were selectively wiped or *Blank Slated*, as is the usual procedure for these highly secret programs. Or perhaps, they have to lie for National Security reasons. Or, it might just not be politically expedient for Obama to admit being involved in the Mars Jump Rooms project. Someday, the truth will come out and we will all know.

One secret space program informant, Bob Dean, revealed on Project Camelot that he had been in a number of massive Deep Underground Military Bases (DUMBs) that could hold 65,000 people each. Also, he revealed that he had been transported in portal systems or modern Star Gate systems, where he had been transported vast interstellar distances in instants of time.

Once, he was in Nevada at Area 51, sub area S-4. There, he entered a building, entered an elevator and was taken 14 levels below ground. At the bottom, he walked out of the elevator and saw a room with a GI sitting at a desk who said, "Welcome Commander, come on in." When Bob Dean entered the room, the GI said, "Welcome to Pine Gap…" Dean asked, "What the hell are you talking about?" The GI replied, "Sir you are in Australia". Dean replied, "No, we're in Nevada." And the young man replied, "Sir, when you just walked through that door, you're in Australia."

So, it seems that they had a pretty seamless teleportation system at S-4. Dean didn't even notice anything walking through the door.

Teleportation can also be done with equipment on board the Solar Warden space ships. This is according to the most informative secret space program whistleblower to date, Corey Goode. His background in the Secret Space program (SSP) and later as a representative in the SSP Alliance is described in The Covert Colonization of Our Solar System. So, I won't repeat all the story here.

Essentially, Corey Goode was found to be an intuitive empath by way of testing in school. At age 6, he was in a MILABS program that enhanced his abilities and later was inducted into the Solar Warden Program. His intuitive empathic abilities were used in extraterrestrial diplomatic meetings to sense if danger was present or deception was being used.

After putting his time of 20 years in solar Warden per the "20 and back" agreement Goode worked back in Texas for a while. But then, he was asked to be a representative between the Sphere Being Alliance and the Secret Space Program Alliance [SSP Alliance]. As Corey Goode explains:

> "The Earth Alliance groups include a significant percentage of the US defense and intelligence community who are fighting for the good. This includes a wide variety of computer hacks that are already exposing hidden truths.

We are also hearing that massive data dumps could occur in the future that will expose the public to Full Disclosure – including the wrongdoings of the Cabal and the truth of the Secret Space Program.

Ultimately the Cabal [New World Order] was created and run by negative extraterrestrials, including a reptilian group known as the Draco. The humans in the Cabal are not shape-shifting reptilians, as biological life does not have this ability. However, certain Cabal members do have telepathic connections to these beings that some intuitives will see as a reptilian overlay on their faces. This accounts for the many cases David Icke has reported on in his earlier works.

The Draco are a conquering race, and Tompkins' new testimony revealed that they contacted the Nazis not just to hopefully subdue the Earth, but to create an army that could be used to conquer many other worlds as well. The Nazis had the mentality and industrial know-how that the Draco wanted, and began colonizing space with the technology the Draco gave them in the late 1930s.

The United States was developing its own space program due to assets it had acquired, but did not succeed to anywhere near the degree that the Germans had. Ultimately the US was bullied and blackmailed into joining the German SSP, and their hope was that they would be able to take the program over and squash the Germans. Unfortunately, the opposite took place – at least for many years.

The level of technology in the SSP dwarfs anything we have on Earth, and includes replicator devices, teleportation, time travel, anti-gravity, free energy and healing technologies that make all forms of disease, illness and even aging obsolete. These technologies would transform our society into a true Star Trek age overnight. These assets already exist in our skies and we simply are not being allowed to see or use them, thanks to the Cabal and their overlords.

However, an ever-increasing column of resistance has developed in the SSP that we are calling the SSP Alliance. Though my formal work in the SSP finished in 1987, I was recently contacted by a new group of powerful extraterrestrials and this led to me being brought back in – upon their own direct request.

This new group was originally called the "Sphere Beings," since they arrived in a magnificent array of gigantic spheres – some of which were as large as the planet Neptune in diameter. Although the first sphere appeared in the 1980s, they didn't begin showing up in larger numbers until the late 1990s. Another major influx appeared right around the Mayan Calendar end-date of 2012. The spheres refused to communicate with the SSP or the SSP Alliance in any way, causing great concern and mystery.

I began working closely with David Wilcock as of October 2014, after our original written contract had started five years earlier. Although David was initially skeptical about my experiences and claims, he soon realized I was validating dozens and dozens of specific data points he had heard from other insiders. I was also quite astonished at how much he knew and had not publicly revealed about the SSP. This information was being held back so he could find out who was real and who was a fraud.

David meticulously wrote down everything I said in real time as I was speaking, which grew into a 150-page, single-spaced document by late February 2015. We are now beginning the process of turning this into a published book and will update you as it progresses.

David was also sharing the contents of this document with key people at the Gaia television network who had earned his trust. David gave an initial "leak" of the data I was giving him, along with data from others, at his Conscious Life Expo talk that same February. This created quite a stir in the insider world and led to me being threatened, though I had not given him most of the intel he shared…

Shortly after we went public with this information, the "sphere beings" finally communicated with the SSP Alliance. They asked for me by name. I was brought up at the beginning of March 2015 and introduced to key people in the Alliance through my contact Lt. Col. Gonzales, who the sphere beings had also been in touch with.

The beings who facilitated this meeting had already contacted me in the past. They were humanlike but also possessed bird-like features, including blue and purple feathers, and thus were called the Blue Avians. The two beings I had the greatest interaction with called themselves Ra-Tear-Eir and Ra-Rain-Eir. The first word "Ra" was pronounced almost like the word "Raw".

The Blue Avians refused to communicate directly with the Alliance, but ended up using me as their messenger in this first meeting. They stood behind me, along with another type of being we simply call the Golden Triangle-Headed Beings, and communicated answers to the Alliance's questions by speaking directly into my mind. I then reported the answers as meticulously as possible.

This was a great surprise, and was only the first of many meetings that followed. I became a vessel of communication for the SSP Alliance and their planet-healing efforts to become widely known to humanity. This effort was greatly assisted with David's cooperation, and even more so when Gaia decided to create an entire television series around these events entitled Cosmic Disclosure." (10)

The sphere beings or the ancient guardians started manifesting their giant planet sized spheres in the 1980s. Pete Peterson revealed that a Neptune-sized object called "The Seeker" came into our solar system in the 1980s, terrifying the Reagan Administration. Many more spheres, usually about the size of the Moon, appeared in the late 1990s and early 2000s. They came in, cloaked themselves, took a position and then just remained there. They did not answer any hailing signals.

There were plenty of visible signs of these arrivals in the "Sun Cruiser" phenomena many observers were catching off of NASA's SOHO satellite photos. At least 100 more spheres arrived around the time of the Mayan Calendar end-date of December 21, 2012. Now, they could be as large as Neptune or Jupiter in size. Again they assumed various positions and then simply remained silent and motionless. The spheres appeared to be intended to help manage the much-anticipated solar events and consciousness shifts expected to occur during this timeframe.

In any case, getting back to teleportation technology, according to Corey Goode, the ship's teleportation can be done without another device on the other end. And this teleportation can be done directly into underground facilities from space without any problems. He himself has been teleported into underground bases, right through solid rock!

Another issue with operating an interstellar fleet, is communication. If two ships were 1 Astronomical Unit (93 million miles) apart, it would take 9 minutes for a normal electromagnetic signal transmitted from one ship to reach the other. So, if you asked a question you would have wait for almost 20 minutes for a reply. In space battle conditions, that is just too long. Normal electromagnetics just won't do.

These ships probably equipped with scalar field or torsion field communicating devices. The Russian scientist Kozyrev has experimentally demonstrated that torsion waves have a speed at least 9 orders of magnitude (10^9) times the speed of light – if not instantaneous! Thomas Bearden has claimed that scalar waves are transmitted along the time axis – not through space. This property of scalar waves also allows them to be transmitted instantly across the vast expanse of interstellar space.

As far as other hardware, the Mars Defense Force Marine special services soldier, Randy Cramer described some of the technology used by the Mars Defense Force.

He described the *Archer Series, Full Body Power Assisted Armor and Environment Suit*. This suit not only provided armor but also was light weight, and had a self-powered exoskeleton that enhanced the soldier's strength and assisted breathing in the thin Martian air.

The helmet for this suit had a heads up display with multi spectral inputs, infra-red, night vision and sonic. This display could detect incoming projectiles from over one thousand yards distance and light them up on the visor. This actually allowed an alert soldier with good reflexes to actually dodge incoming bullets. This secret technology is far in advance of what the marines are equipped with here on Earth.

Randy Cramer was blown up in battle many times. He explains that for example, if a soldier had his arm blown off above the elbow, the highly trained medics would use what Randy Cramer called "a high definition holographic cellular regenerative technology machine", to totally regrow a new arm in 5 hours in the hospital. Amazingly, the soldier would be ready for battle the next day. His and his comrade's lives were saved many times with this advanced medical technology.

Corey Goode states that his meals were created in a replicator they called a "printer" These replicators could create a meal out of raw material like coconut oil or hemp oil. These meals would taste pretty much like the real thing. Corey's favorite was roast beef.

So, here we see that much of the science fiction series *Star Trek* seems to be based on what was actually being planned for the secret Solar Warden program and the ICC enterprise.

The *Star Trek* Series started in 1966 - the same time frame that this technology was being developed in secret think tanks in California and elsewhere.

According to Tom Seldon, one of the production assistants of *The Outer Limits*:

> "Star Trek was in fact an outgrowth of The Outer Limits. Gene Roddenberry watched our dailies all the time and took a lot of phone calls from our screening room. He was spurring his imagination and checking on the incredible quality control we had. I wondered why he was there but he was there more often than not during the time he was coming up with Star Trek."

The *Outer Limits* was created by Leslie Clark Stevens IV. The following firmly suggests that Roddenberry and Leslie Stevens IV had reached a business arrangement for the planned sci-fi series, *Star Trek*:

Bearing in mind that Roddenberry was contracted to a rival studio and a rival network, the odds are essentially slim to none that the two men didn't have some kind of business arrangement, whether in writing or not.

If Stevens and Roddenberry had indeed developed a business arrangement for the new *Star Trek* series, this is where Stevens background becomes critical in understanding the nature of their arrangement. Stevens was the son of a U.S. Navy Vice Admiral Leslie Stevens III, who died in 1956.

Vice Admiral Stevens was a contemporary of Rear Admiral Rico Botta, who oversaw the covert navy spy program out of Nazi Germany to learn about Nazi flying saucers during World War II. The 29 Navy spies in the program had not only learned that the Nazis had developed up to 30 different flying saucer prototypes, but were also being directly being assisted by the Draco reptilians in building secret bases in Antarctica.

Both Admirals Botta and Stevens were leading experts in aerospace engineering and headed top Navy aerospace facilities at various points during their careers. In 1946, both were retroactively promoted to the rank of Rear Admiral as of 1943, for their wartime duties.

Botta went on to head the Naval Air Material Center out of Philadelphia Shipyard from 1950 to 1952. It is from this and similar command assignments that plans began for a secret Navy space fleet aimed to counter what the Nazis had developed in Antarctica.

Similarly, Admiral Stevens was known for his accomplishments in aeronautics had a hand in the design or conception of all naval aircraft, aircraft carriers and carrier landing apparatus. Admiral Stevens' aeronautics expertise meant that he was almost certainly aware of what Botta had learned about Nazi aerospace projects.

It is more than likely that Admirals Botta and Stevens were the first to comprise a covert Navy leadership group that would oversee the development of a U.S. Navy space fleet based on modern aircraft carrier battle groups that would in time join an interplanetary alliance. It is a good possibility that Rear Admiral, Leslie Stevens III, informed his son of the planned Navy Space fleet. And Leslie Stevens IV fed this information to Roddenberry for his *Star Trek* series.

As a matter of fact, David Wilcox in his new book, *The Ascension Mystery,* states that many science fiction movies are based on real things going on in these secret space programs. This is done for several reasons.

One, if any disclosure from an insider occurred, it could be said that these people were just copying something from the Sci-Fi movie. Sometimes, the movie title would even be the actual secret project name, like *Ice Station Zebra*.

Two, it provides a means of showing off what the elites actually have to the public unconscious mind without actually admitting to anything. This provides a subconscious programming method to engender greater authority worship and a feeling of helplessness among the masses.

There are many at the Pentagon that want the UFO and secret space program secrecy to end and are part of the secret Earth Alliance working towards disclosure. They have to be careful however, because they understand that whistleblowing can be lethal. Also, these military are "outside the loop" on higher levels of secrecy.

Corey Goode makes it clear there are two levels of secrecy in the secret space programs (SSP). The lower level – the military space program - has no knowledge of the joint E.T. - secret government alliances and believes that all UFOs are made here on Earth. They also believe that we haven't traveled beyond our solar system.

The higher level SSP knows all about interstellar travel and commerce with ETs. These more secret programs are controlled by certain corporations, like the Interplanetary Corporate Conglomerate (ICC), certain Illuminati families and deeply covert sections of the CIA which oversaw the former MJ-12.

The *Stargate SG-1* T.V. series depicts some of the actual portal technology and precautions being used in actual secret programs that can travel to distant star systems almost instantaneously. Insider, Daniel informed David Wilcox that this series was chock full of truthful information that he had encountered on his job.

According to Corey Goode, there are natural portals connecting the 38 different stars in our local star cluster, which our sun happens to be in the center of.

As discussed in *The Covert Colonization of our Solar System*, these portals are created by torsion fields from massive spinning bodies. Since all these planetary and solar bodies are in constant motion, accessing these portals have to be done at the correct time as well as place. All that is part of the Star Gate addressing algorithm. The secret of doing this was originally given to the Vril society in Germany by the ETs from Aldebaran in the Taurus constellation and later to our secret government from other ETs.

Also, our sun has a natural inter galactic portal. Consequently, a much higher than normal amount of ET traffic occurs in our solar system. (12)

Another issue with the ET presence around our planet is understanding their agendas. One large agenda that was made self-evident early on after the event of the nuclear age, was their concern that mankind would destroy the planet in a nuclear holocaust. The technology of Earthmen was way ahead of their wisdom to use it sensibly or humanely, as witnessed at Hiroshima and Nagasaki.

This agenda of one segment of our ET visitors was soon made manifest by the large number of UFO sightings around nuclear armed ICBM silos and nuclear test sight sites. As early as 1945 many UFOs were sighted around the White Sands testing ground when the world's first nuclear bomb was tested.

The ICBM installations would frequently be put out of commission and rendered incapable of executing a launch by these UFOs. In some cases, the nuclear warheads were melted down and made nonfunctioning. Test missiles launched from Vandenberg were occasionally shot down by UFOs

Apollo Astronaut, Dr. Edgar Mitchell, who had spoken with a number of personnel manning these facilities, verified the above information. Mitchell stated, "My own experience talking to people has made it clear the ETs had been attempting to keep us from going to war and help create peace on Earth."

On September 27, 2010, Robert Hastings held an event at the National Press Club in which seven Air Force veterans shared their testimony about these encounters. Hastings claims to have interview over 150 military veterans who were involved with various UFO- related incidents at U.S. missile sites, weapons storage facilities, and nuclear bomb test ranges. Hasting contends that these ET beings are occasionally disrupting our nukes to send a message to the American and Soviet/Russian governments that their possession and potential large scale use of nuclear weapons threaten the future of humanity and the environmental integrity of the planet!

Shortly after Hasting's National Press Club event, on October 23, 2010, a cigar shaped UFO appeared over, F. E. Warren AFB in Cheyenne, Wyoming, the largest nuclear missile facility in the U.S. This base lost its ability to communicate with 50 of its Minuteman III missiles, representing one ninth of the entire U.S. arsenal, when this happened.

So, these ETs are watching out for humanity and have the will and ability to prevent a nuclear holocaust if some idiot in charge ever decides to start a nuclear war.

Colonizing the Solar System

It is important to understand that the Nazis had treaties with the Draco Reptilians which allowed them to have a base on the Moon. This information is verified by both William Tompkins, who knew that the Dracos and Nazis had treaty agreements and Corey Good, who knew that these agreements included a base on the Moon.

Corey Good explains that the Moon is a diplomatic zone for many ET civilizations. Each have their own sectors and have a strict code of conduct. Even if the different civilizations may be at war elsewhere in the galaxy, they might have bases within kilometers of each other on the Moon and must peacefully co-exist on the Moon.

After the secret treaty between U.S. President Eisenhower and the Nazi International which turned the Military Industrial Complex (MIC) over to the Nazi control, the U.S. was invited by the Nazis up to their moon base. After the Nazis gained full control of the U.S. Military Industrial Complex in the late 1950s, the LOC was greatly expanded to its present size and was eventually turned over to the U.S. secret government and called the Lunar Operation Command (LOC). The Lunar Operations

Command (LOC) controlled all Lunar and near Earth space missions.

So even though the Tompkins planned a 2,000-man Navy Moon base, the secret NOVA program was not allowed to be built by the ETs on the Moon, the U.S. eventually gained access to the LOC because of the U.S. - Nazi secret alliance and the Nazi - Draco alliance.

The Nazis then built another base in another Draco sector of the Moon from where they secretly operate their *"Dark Fleet"*. The *Solar Warden* forces have little knowledge of the, very secretive, *Dark Fleet* which operates primarily outside of the solar system. This base is partly above the surface and has a pentagon shape.

In the 1980s, the Nazi Controlled U.S. Military Industrial Complex (MIC) had created sufficient Navy space ships and bases on the Moon and Mars and elsewhere, to police the Solar System in a program officially called *Solar Warden*. After a level of security was established, the commercial aspects of this technology were pursued. A space based economy could provide financing for *Solar Warden* to make it self-financed and also generate profit for corporations.

The idea was that *Solar Warden* provided the Solar System with security to allow trade with friendly extraterrestrial civilizations without the risks of being raided by hostile ET groups. A conglomerate of large corporations on this planet secretly created an interplanetary one, called the Interplanetary Corporate Conglomerate (ICC). Subsidiaries of the ICC were also formed, one being the Mars Colony Corporation. CEOs from known Earth based multinational corporations are selected to become CEOs of the ICC.

This information comes from Corey Goode. He spent many years on a *Solar Warden* ASSR "ISRV" (Auxiliary Specialized Space Research, Interstellar Research Vessel) the *Arnold Sommerfeld*. He and all others in the "20 and back program" were totally cut off from contact with Earth. To keep boredom away, he would spend much of his free time studying his smart glass pad.

This pad had much of the real history of the Earth, Solar system and the secret space program included. He was allowed access to this information because of his diplomatic duties as an Intuitive Empath. He needed good background information on subjects discussed at these ET diplomatic meetings to do his job.

His access to the information on these smart glass pads and these diplomatic conferences made Corey Goode one of the most informed of the secret space program whistleblowers so far.

Other sources of this information include Vladimir Terziski, a student of the Russian Academy of Science, who did considerable research into the World War II Nazi Space program in Eastern Europe after the Collapse of the Soviet Union. He has provided me and others with a lot of information on the Nazi flying saucer projects.

Solar Warden has sub operations called the *Earth Defense Force* and the *Mars Defense Force*. *The Mars Defense Force* is contracted by the Mars Colony Corporation to defend its assets on Mars.

This leads to the sorry situation that the CEO of the Mars Colony Corporation can give orders to the commander of the *Mars Defense Force*. This led to a disastrous decision to raid a Reptilian Temple after a peace treaty between the Reptilians and Mars Colony Corporation had been agreed to - because a greedy CEO wanted some artifacts at the temple!

The *Mars Defense Force* commander knew it was a terrible idea for a number of reasons, but was outranked. The resulting raid was a disaster. Of the original 976-man, four division combat team only 35 survived the raid. Nearly four divisions had been completely wiped out. Randy Cramer was one survivor to tell the story, which is related in *Secret Science and the Secret Space Program* and elsewhere.

Another ugly feature of the Mars Colony Corporation is the use of virtual slave labor. The Nazis used slave labor in their armament factories and underground facilities during World War II and, through the ICC, are still using it on Mars to this day.

The colonists, many of high intellect, were brought to Mars starting in the late 1960s. They were told that a global disaster would soon happen to the Earth and they were needed on Mars to preserve the human race. This caused what later was called the "Brain Drain" in England. But, it was also happening in other countries including the U.S. It was an *Alternative Three* scenario. *Alternative Three* is a science fiction story based on what was secretly going on. (11)

Only persons of the proper IQ and Racial features were selected in this colonization program in typical Nazi fashion. And, everything was done in complete secrecy.

Typically, a bunch of postcards were written up by the prospective colonist before leaving Earth. They would claim on these postcards that they had received a wonderful job in a foreign country. These postcards would be gradually post marked in the different country and sent off to their relatives by agents of the secret recruitment program.

So, the relatives, after reading these postcards, would think the persons were in a different country enjoying their new job, and not worry too much about them. After a few years, they would mostly be forgotten about.

Once on Mars, they were put to work, first building the colonies which were mostly underground, and later worked in the factories to produce trade goods. They were cut off from all news or communication with Earth. They were later told that a cataclysm had actually occurred on Earth that wiped out all life on the planet and that they would never be able to return.

The colonists were paid with wages that provided the basic essentials to live. The wages were paid back out to company stores. So, it wasn't actual slavery since they were getting paid. But, it was the next closest thing. The colonists were basically owned by the Mars Colony Corporation and had no chance of ever leaving.

The Mars Defense Force had their bases of operation separate from the colonies that they were defending. No contact was allowed between soldiers and the colonists. This was because the soldiers came from Earth and knew that no cataclysm had actually occurred. This was according to both Randy Cramer of the *Mars Defense Force* and Corey Goode of *Solar Warden*. Goode also states that he knew of five different colonies on Mars. Goode also explained further:

> "The ICC has an entire industrial infrastructure that includes bases, stations, outposts, mining operations and facilities on Mars, various moons and spread throughout

the main Asteroid Belt (where a "Super Earth Planet" once existed). They have facilities to take raw materials and turn them into usable materials to produce both complex metals and composite materials that our material sciences have not dreamt of yet. They have separate groups of facilities that produce various types of technologies as well as each facility or plant that produces a specific component of a technology so that those working in the facilities and living in the support colonies/bases do not know exactly what they are producing. Much of the time the components are multiuse and are used in cross over projects. There are facilities on Earth that operate in much the same manner that contribute to the SSP on several levels.

There are other bases on Mars that are controlled by Military/Security groups as well as some scientific outposts. These can be owned and maintained by other SSP Programs but are usually going to report to the ICC on some level since the ICC controls much of the Air Space and Security Operations on and around Mars. Most of the security personnel that are assigned to Mars are assigned to and serve under the ICC. The military groups that will be returning to their previous organizations (SSP Groups) are kept isolated from the population and personnel who live and work on the Colonies, Bases and Industrial Facilities that they protect. They are normally in the rather Spartan outposts that I have described previously in other writings. I had seen a few of these outposts built from the "Ground Up". They were always quite a distance from the main underground colonies, bases and industrial facilities and spread out in a Multi-Teared Perimeter Defensive type of system. There are "Non-Humans" also having bases on the planet. Some of them have been there for some time and have the highly coveted larger lava tube systems that have been built out into base systems that are unimaginably huge and can securely reside millions of inhabitants."

The secrecy around these operations is nothing short of astounding. Men who sign up for these defense forces are required to sign up for a 20-year tour of duty. When this time is served the men or women are returned to the Lunar Operation Command. There, they then have their memories selectively erased of their entire 20 years of duty. Then, they are age regressed by 20 years which requires about 2 weeks with advanced medical technology. Then, they are sent back 20 years in time to the time starting the tour of duty. This policy was called "20 and back". Theoretically they would never know that they had done this tour of duty and their lives could normally continue without revealing any secrets.

When Corey Goode was asked what happens when you got back. Did you remember things? Are a lot of people out there right now who served in these programs but can't remember anything? Corey Good replied:

> "Well, yeah. There are many, many, many thousands of people. And many of them are programmed to be triggered by this topic. And some of the biggest skeptics are people that actually participated in this. And, yes, for the most part, the blank slating is very effective. And 3% to 5% of the people, the mind wipes or the blank slating does not work. And it's usually the people that are intuitive empaths that it doesn't work in, because they are not just relying on their memories on their hard drive in their brain. They also have a strong connection with their higher self - sort of like a virtual hard drive - with their

light body and higher self. Like, you know the same way people have memories of a prior life? Well, in the birthing process in this chemical brain of ours, you can't possibly remember a life 1,000 years ago. But in your soul memory, you can. So in these programs, they can erase these chemical memories and electromagnetic memories in your mind and play with them. But they can't affect the memories in your soul body or in your higher self. So in the 3% to 5% of people that have that connection, they have a very difficult time of affecting those people with screen memories and blank slating"

According to Andrew Basiago, during the 1980s, he had gone to Mars a number of times from a teleportation chamber called a Jump Room in El Segundo near Los Angeles Airport. A Number of others in the same program have also come forward and said that they had also gone to Mars via the El Segundo Jump Room.

Michael Relfe, a member of the *Mars Defense Force* also speaks of this Jump Room. VIPs would come to Mars via these Jump Rooms and visit for a period of time, usually less than 2 weeks, and then return to Earth the same way This is according to the *Mars Records* that his wife composed based on Michael Relfe's recovered memories. Relfe and other grunts in the *Mars Defense Force* on the other hand would have to be subjected to the "20 and back." policy. (13)

Another person, Tony Rodrigues, has recently come forward who was recruited into the "20 and back" program.

Tony Rodrigues claims that in 1981, when only nine years old and in 4th grade, he was involuntarily recruited into a "twenty and back" program as punishment for something he did to one of his class room peers.

He publicly embarrassed the son of a high level member of the Illuminati and Rodrigues says that he was soon after abducted by five aliens. He was then genetically tested to determine what skills he possessed, which could be used in covert 'support' programs and eventually for the secret space programs themselves when he got older.

Rodrigues says that he was first forced to work as a psychic for a drug running operation out of Peru for four years. He would be put asleep with drugs and would psychically determine if a particular drug run would be safe or not. Edgar Casey, the sleeping prophet, also was a very accurate psychic that worked while unconscious.

Later he was recruited as an underage sex worker in the area of Seattle from age 13. His wealthy Illuminati owner was a practicing Satanist who hosted group sex orgies and even ritually sacrificed a baby. Rodrigues was actually forced to eat parts of the dead baby.

When he turned 16 years old, in 1988, he was taken to the Moon to be tested for any skills he possessed that would be of most benefit for service in the secret space programs. He was taken to both the Lunar Operations Command or LOC and the pentagonal/trapezoidal shaped Nazi Dark Fleet base, which he observed before landing.

While briefly serving on Mars, he was attacked by a large insectoid and had both his arm and foot bitten completely off. He applied tourniquets above the wounds and was carried by a Mars Defense Force fighter to the base hospital and passed out. The next morning, he woke up and discovered that he had a new arm and foot and was completely healed!

That hospital technology seemed to be similar to what was described by Randy Cramer, where MDF soldiers blown up in battle could be completely repaired and rejuvenated overnight and ready for battle the next day!

After this incident the project, which Tony Rodrigues was involved with – basically acting as bait to draw the insectoids in so that the Marine special services forces could kill them – was abandoned. At the outpost, the Marines s.s Mars Defense soldiers would always playing pranks on and teasing members of his group.

Later, Rodrigues was taken to *Aries Prime* and tested to see what abilities he possessed. Then, he was shipped out to Ceres, a planetoid in the asteroid belt, where he worked for 13 years with a German led freighter crew that was part of the "Dark Fleet," initially set up by Nazi Germany during World War II.

When he got off the flight to Ceres and entered the hangar with others who had accompanied him, their group was informed that they were required to obey every order given them. They were told that a few others before them that didn't agree to obey, were shot on the spot. Then, they were asked if any of them needed to be shot. Needless to say, none of that group thought so.

The operation there was very much a Nazi operation. There were swastikas everywhere. Everyone had to give the stiff arm salute and say Hiel - followed by the name of the leader of the Ceres operation, which would change every so many years. German and English was the common language of the base. But, there were also Extraterrestrials on the base. Also, there was a percentage of women present. Many of the permanent members of Ceres were born there and would probably die there.

Many of the laborers there were mining the mineral resources of Ceres. When some areas were mined out, cities would be built inside the resulting caverns. Others were involved with manufacturing. And trade was carried out with other civilizations both in our solar system and other star systems.

Rodrigues was assigned to ship repair, at first mainly in plumbing, replacing old failed valves. The older ship interior looked like a submarine, with valves and gauges everywhere. Valves and gauges were not threaded but welded into place. Fortunately, he did have automated cutting and welding tools, which made the job easier and produced welds which could pass inspection.

Eventually the old ship was replaced with a newer model and Rodrigues was promoted to the cargo section. Even though his life became improved, he like others of his rank, still had to always wear a collar with which he could be electro shocked if his Nazi superiors thought he needed to be disciplined. Now, he actually would fly with the ship to different destinations and deliver the cargo.

Later, he would visit a base brothel. The women who worked there were young and pretty and were all mind controlled. Before being brought there, their memories were totally wiped out or Blank Slated. So, having no prior memories of their childhood or families, they believed everything they were told. In addition to the men on the Ceres base, many aliens would also visit the brothel and have sex with these unfortunate women.

The Nazi women and men who lived there permanently, always had a superior attitude because of their advanced technology and tended to bully people like Rodrigues. These people had to just accept their fate and wait out their 20-year tour of duty, because escape was impossible.

This is merely a short synopsis of the Tony Rodrigues story. He is planning on writing a much more detailed story in book form. The first segment of five planned interviews with Dr. Michael Salla is here: http://exopolitics.org/20-years-a-slave-in-secretspace-programs-abduction-programming/

So here we see the gross unfairness in the way corporations are operated. Here on Earth, the CEO's of these corporations receive thousands of times the salaries of the employees. Are they actually worth that much more? It is highly doubtful.

These CEO's have been streamlining their company by firing much of the corporate staff and employees while making the remaining staff and employees work twice as hard. These tactics increase the corporate profit margin which keep the stock prices elevated. But, the company shareholders usually don't see any increase in dividends from these actions because the CEOs give themselves even more pay and benefits like stock options and golden parachutes!

As we can see, the Mars Colony Corporation have a captive work force and the corporate unfairness is even much greater on Mars, where no government regulations are in effect to protect worker's rights. The workers on Mars suffer from a similar plight that the negros in the Southern plantations before the Civil War suffered, but in a more modern context.

The trend on this planet unfortunately is heading in the same direction. In the last part of the 20th and early part of the 21st century, we have seen a number of international trade treaties being formed, Like the WTO and NAFTA which removed many jobs from the U.S. where people received a decent wage and moved these jobs to countries where basically slave wages were paid.

Now the Trans-Pacific Trade Partnership (TTP) is being pushed forward. This treaty would give corporations more power than governments. They could sue governments if they lost revenue because of a government regulation. So, basically they could pollute at will, and treat their workers like slaves and be above all government regulation. Profit seems more important than life itself according to the corporate CEO's way of thinking.

Despite the compartmentalized secrecy, many in the Solar Warden program became aware of the gross unfairness and inhumaneness of the way the secret colonies in our solar system were being run by the ICC. Many of these military persons were from the U.S. and took an oath to uphold the U.S. Constitution and strongly felt that human rights should be upheld.

Eventually, there was a well-organized revolt by commanders within the Solar Warden fleet and a forming of a covert alliance to counter this corporate fascism. The alliance was called the Secret Space Program Alliance or SSP Alliance.

Although many in the SSP Alliance lived off world or in their space craft, some of the persons in this Alliance had family and homes back on Earth. These members of the SSP Alliance had to operate undercover for their own personal safety and that of their family, when they returned back on Earth.

Also, an Earth Alliance was formed to counter the Cabal surrounding the New World Order, a fascist project conceived by the Vatican, as documented in *The Secret History of the New World Order*. Most members of the Earth Alliance also have to operate undercover for their own personal and family safety.

Both the SSP Alliance and the Earth Alliance have undertaken to bring down the Cabal and restore true freedom, without economic bondage, to the earth and the colonies of our solar system. One outward manifestation of their secret endeavors is the BRIC financial organization.

This is the real reason for the reemergence of the "Cold War" between Russia and the Western powers. Russia is definitely part of the Earth Alliance, while the Cabal are still the controllers of the West.

Before long, a limited war in space, broke out between the SSP Alliance and the ICC. In the process, an entire Mars colony was accidentally destroyed as "collateral damage" by the SSP Alliance. Many of the SSP Alliance felt pretty badly about this incident, as many innocents were killed. And, at the same time, the CEOs of ICC began to worry about losing all their colonies to these "rebel forces."

Before long, a ceasefire and negotiated truce was called for. It seemed that military violence was not the proper way to solve the problems of human selfishness and unfairness. It is a hard lesson for humankind and their leaders to learn.

The SSP Alliance was charging the ICC with crimes against humanity because of the conditions of the Mars colony workers. The ICC was denying that slave labor was being used. Part of the negotiations called for representatives of the SSP Alliance to inspect a Mars Colony of their choice.

Corey Goode was chosen to be one of these SSP Alliance representatives, by the Sphere Being Alliance.

This was not what the Sphere Beings called themselves and some posit that they are the ancient guardians or builder race of this portion of the galaxy. They made their reappearance before 2013 by placing huge planet sized spheres around our solar system and the solar systems of our nearest star cluster. These spheres are mostly invisible, but some close to the sun have been videoed by NASA. https://www.youtube.com/watch?v=bQ7RaOMHb5I

The purpose of these cloaked spheres is to moderate the galactic waves of energy, capable of causing planetary upheavals, now impinging our part of the galaxy from the galactic center. This action to moderate global disasters seems part of their role as guardians.

Many in the ICC tried communication with this alliance but received no response. So they just called them the Sphere Being Alliance.

However, the Sphere Being Alliance did communicate with the SSP Alliance. And they choose Corey Good and a person we shall call Gonzales (he needs his real name kept secret) to be their representatives.

In any case, Corey Goode was called on to be part of the Mars Colony Inspection Team.

An account of their inspection of Mars colony conditions, subsequent arrest and rescue by the Sphere being alliance, as described by Corey Goode, is described in *The Covert Colonization of our Solar System* and here: http://spherebeingalliance.com/blog/joint-ssp-sphere-alliance-icc-leadership-conference-tour-of-mars-colony-on-6-20.html

The Lunar Operations Command (LOC) was still under the Control of the ICC after the formation of the SSP Alliance. And the LOC, under orders from the ICC, was preventing known SSP

Alliance space ships from returning to Earth.

Corey Goode was being transported to his various meetings via the Sphere Being Alliance in a way that the LOC had no control over. This method is described as the sudden appearance of a blue orb about the size of a soccer ball that would hover in the air and dance around in front of Corey Goode. Goode would telepathically let the Orb know that he was ready. Then the Orb would expand and place Corey inside itself. The orb would then quickly transport Corey Goode where he was supposed to go anywhere in the solar system.

Corey Goode explains, "It's not a technology. It was explained that they are the highest density of the Sphere Being Alliance. They said they're said to be ninth density beings that appear as orbs or spheres. And the size does not matter. I mean, it can be the size of a ping-pong ball, or it can be so large that it can contain an entire solar system."

After the Mars colony inspection team, that included Corey Goode, was arrested in the inspected Mars Colony in the Martian Southern hemisphere and faced probable death, they were rescued from their Mars colony prison cells in this manner by the Sphere Being Alliance.

There many factions in this secret space program just as there are many factions vying for power among the secret cabals here on Earth. Also there are other secret societies which are fighting the New World Order fascist faction. Benjamin Fulford speaks of the Oriental White Dragon Society which is very much opposed to the present unfair global financial system and those syndicates like the Vatican, Black Nobility, Rothschilds, and Rockefellers in back of it.

Leaders in the U.S. Military also are aware of the criminal factions like the Bush and Clinton factions that have led the U.S. into so many unconstitutional, unnecessary, overt and covert wars which caused so many unnecessary casualties among our troops and ruined the U.S. reputation at home and abroad.

There is a real possibility of a military coup in the U.S. just like the rebellion in the Solar Warden Program. This coup would be designed to restore the Constitution and arrest high level criminals in government that presently think that they can literally "get away with murder." The U.S. could then have Nuremberg type war crimes trials against U.S. war criminals without being militarily defeated first.

In any case, there is a lot of secret infighting going on behind the scenes which the controlled news media is not covering - both on the planet and out in space.

Another factor adding to the confusion is the high level of compartmentalization in the secret space program. Many in the "lower level secret space program," run by the NRO/USAF/DIA. had no idea of *Solar Warden* since they did not possess "a need to know."

This is how Corey Goode explains this compartmentalization while being asked about the past USAF announced the Manned Orbiting Laboratory (MOL):

> "The Earth Based Military Black Ops Services have had separate and ongoing lower level space programs from the beginning. They are very compartmentalized, and those who actually operate craft and stations in various orbital zones around Earth never have an opportunity to see what else is going on out there. NASA ISS see their

activities, these lower level military black op's programs see the NASA, Advanced (Breakaway) Secret Space Program activities as well as some activities of off-world groups. They know there is a lot more going on beyond their programs but do not have a need to know and are debriefed routinely by the Earth military forces to get as much information as they can about what may be going on in the Break Away Secret Space Programs that are more advanced. Very few below those in command of these black op's programs and the operators know of the other activities going on in the Sol System. The support and infrastructure portion of these black op's programs are not read into the small circle of information known by the operators and command structure and are kept oblivious of the other activities being observed outside the atmosphere of the Earth.

That being said, there have been many other space stations and similar training labs that have been launched into orbit for periods of time, used in secret and then allowed to burn up in reentry after they have served their purpose. I do not know if the particular orbital laboratory[MOL] was ever an active program or not. The LOC [Lunar Operations Command] would keep up with the "traffic" of all space programs (civilian/military/intelligence satellites, NASA, secret military, other nations and of all of the Advanced Secret Space Programs) and their assets to try to plan and schedule their locations to prevent them from being visible to each other as much as possible. This includes keeping track of ground based astronomical observatories and their highly documented and controlled access times and the windows of the sky they are scheduled to observe. This is why there are no real accidents when it comes to SSP vessels flying through the live video feeds of the NASA ISS. The traffic is very highly planned and tracked and if a vessel/craft is caught on the live feed it was done on purpose."

Some of the intelligence released by Corey Goode, over the last year and a half, about secret space programs, was accurate enough to come to the attention of the lower level secret space program. Goode stated that he experienced a total of three different military abductions on a spacecraft that was clearly less advanced than those belonging to the SSP Alliance. The abductions happened during the period from January to February 2016. The military abductors were trying to discover if Goode was part of an Unacknowledged Special Access Program and the identity of leaders in the SSP Alliance.

He was heavily sedated with a chemical cocktail fed into him intravenously. Then, he was subjected to an advanced interrogation, using computer aids. For example, hundreds of military academy pictures of suspected military leaders in the SSP Alliance were shown him and an involuntary reflex in the iris of his eyes upon recognition of the person would be recorded by the computer. In this way, he involuntarily "outed" 3 of his compatriots in the SSP Alliance, without even realizing it!

After these procedures Goode had his memory of the abduction erased or "Blank Slated" and returned to home. For a while he did not even realize that he had been abducted. But within months, he recovered these memories. Here is how Corey explains what happened:

> "A tablet with a camera was held in front of my face and academy type military photos were shown to me. The camera monitored my eyes and marked a photograph when it was detected that I recognized the person.

This incident caused the outing of 3 high ranking SSP Alliance individuals and caused a further rift between myself, Gonzales and the SSP Alliance. Because of the chemical interrogation and the attempted blank slating of my memories of the incidents, I didn't remember the full details until I was informed later of the security breach."

This "outing" placed his compatriots into greater danger and meant that they could not return safely to Earth. One of the outed SSP Alliance individuals has subsequently disappeared, another has been captured by the "lower level SSP", and the third was Gonzales – Goode's primary contact with the SSP Alliance.

Due to his involuntary 'outing' by Goode, Gonzales had to leave his civilian "cover" identity in Texas, and quickly found refuge with one of the Inner Earth civilizations called the 'Anshar', which he and Goode had previously met with on several occasions. More can be learned about the Anshar here: http://exopolitics.org/inner-earth-beings-take-first-step-to-openly-reveal-themselves-to-humanity/

The problem was that, in spite of being a good man, Gonzales, still had too much ego to blend in well with people of Anshar and, before too long, he had worn out his welcome there. So the Anshar people took him to a base in the Kuiper belt region of our Solar system where he now resides, unable to return to the Earth and live safely.

It is important to note that Corey Goode was no longer in the *Solar Warden* program at this time but was only acting as a representative between the SSP alliance and the ICC who was selected by the Sphere Being Alliance. As such, he had the protection of Sphere Being Alliance and if he had been truly in danger during these abductions he would have been immediately rescued by them, just as he had been rescued when arrested while inspecting a Mars colony.

The whole affair badly affected the former good relationship between Gonzales and Corey Goode.

Gonzales didn't seem to understand that his "outing" and those of his comrades by Corey Goode had been involuntary, while Goode was heavily drugged. And the information was not given voluntarily – but by the human body's own reflex actions. Gonzales was unable to forgive Corey Goode and severely restricted information was given to Goode in all later meetings.

If Goode had not been chosen representative by the Sphere Being Alliance between them and the SSP Alliance, it is doubtful that Gonzales would have ever met with Goode again.

Current Events on Earth

On March 24, 2016, U.S. President, Barack Obama visited the unofficial Nazi International headquarters in San Carlos de San Bariloche, Argentina, the same town that Adolf and Eva Hitler and Adolf Hitler Jr. fled to in the latter part of World War II, President Eisenhower visited in February 29, 1960 and President William Clinton visited on October 18, 1997.

There, Obama visited the newly elected, pro U.S., Argentine President, Mauricio Macri. This was the first visit to Argentina by an U.S. President in nearly 20 years. It was on the 40th anniversary of the 1976 U.S. orchestrated military coup that placed a dictator in power in that country. Hence, there were many anti - U.S. demonstrations because of this visit on this particular date. Police held back protesters who gathered on the road alongside the motorcade route of President Barack Obama and his family during motorcade's drive from San Carlos de San Bariloche, Argentina to Buenas Aries.

In February 1960, President Eisenhower had traveled to San Carlos de San Bariloche where he officially negotiated the Joint Declaration of Bariloche with Argentinian President, Frondizi concerning Peace and Freedom in the Americas. The real topic of negotiations, however, concerned deals which would further put the U.S. Military Industrial Complex firmly under the control of the Nazi International. San Carlos de San Bariloche was then, the headquarters for the Nazi International and Adolf Hitler and his wife Eva, had been living in the nearby Inalco estates.

This agreement led to the emergence of the Interplanetary Corporate Conglomerate (ICC), the megacorporation colonizing Mars, mining the asteroids and industrializing other planets and their moons in our solar system, as revealed by Corey Goode and others, The ICC also was now in charge of the Nazi bases in Antarctica.

Therefore, it is more than likely that President Obama's visit to Bariloche, was to finalize new deals with the ICC/Nazis, which would facilitate their desire to move large number of people and cargo to safe locations in South America and Antarctica. Obama, in addition to being U.S. President, was also a deep cover CIA agent and insider. (14)

There seemed to be a large movement of powerful individuals into South America around the same time. Reports came in that high level syndicate groups were moving huge amounts of personal items and supplies to South American underground bases most noted in Brazil. More recent reports stated actual family members and high ranking syndicate members were pouring into these underground bases like ants before a storm.

It seemed that what they feared was huge solar storms predicted to hit the Earth. The so-called "solar kill shot" of huge solar coronal mass ejections, long predicted by the remote viewer, Ed Dames, which was imminent according to him.

Corey Goode's May 14, 2016 update went on to describe huge

submarines used to transport people and cargo to Antarctica as revealed to him by Lt. Col. Gonzales, who also confirmed that these people and supplies were in many cases being transported to Antarctica via "Black Submarines" that were EM Driven and the size of container ships.

Also, they were underwater tunnel systems connected to underground bases, built in huge caverns, throughout South America that these subs could take these passengers and supplies to.

Corey Goode stated that there were six large industrial complexes located in the Western Antarctic region. The two largest complexes were city-sized and about two miles wide.

Further, he stated that he saw a secret Antarctic city while traveling on an Anshar space craft that flew about 60 feet above the surface of the ocean directly into what looked like an Antarctic ice cliff, but which in reality was a hologram hiding the entrance to a huge ice cave. The entrance was a huge archway that was over one hundred feet high and led into a huge ice tunnel. Steam or fog seemed to be coming from the entrance.

They flew along for a while with water below and ice overhead. Then, they spotted a small island below with some buildings on it which appeared to be some sort of outpost. Then, they were flying over land. There were pools of water below that had steam coming out of them.

Goode saw large buildings, and two huge black submarines used to transport people and cargo to Antarctica being unloaded by giant cranes. The whole area was lighted up by light from the buildings reflecting off the ice cave ceiling.

All of this was inside a huge ice cave which was created by

geothermal activity. Parts of this cave were around 600 feet high and other parts were about 200 feet high. In addition to buildings and a port inside this cave, a fleet of triangular space craft was seen parked below.

Next, the Anshar craft went under water and followed some transport submarines, through another huge arch, which seemed as if it was built by some ancient civilization. It had suffered the ravages of time and was cracked with several pieces of it broken off. This huge underwater tunnel lead to the open sea. (15)

These large underground and under ice industrial complexes had first been established in the 1930's by Nazi Germany, according to Goode, and subsequently expanded in the 1950's after agreements had been reached with the Eisenhower administration and the U.S. Military Industrial Complex. Currently, these six Antarctica bases are used by the Interplanetary Corporate Conglomerate (ICC).

Goode then described in his May 14, 2016 update, a battle that had taken place over Antarctica.

One of the most interesting things that came out of this briefing by Gonzales was that there had recently been reports of 6 large black teardrop shaped cruisers that belonged to the Nazi/ Draco *Dark Fleet* that were in the process of leaving the atmosphere after breaking the surface of the ocean near the coast of Antarctica.

This secretive *Dark Fleet* is operated by the Nazis in alliance with the Dracos. They have a Moon Base separate from the LOC on the dark side of the Moon. It has a pentagon shaped structure on the lunar surface. It is so secretive that the ICC and SSP Alliance know very little about them. The *Dark Fleet* operates mostly outside our solar system.

Dozens of chevron shaped craft swarmed these *Dark Fleet* cruisers and attacked the leading two craft causing massive and shocking damage. The cruisers broke off their attempts to leave orbit returning to below the surface of the ocean where they came from.

Events in Antarctica were being monitored closely by different nations and/or space programs. Goode described huge spherical craft over Antarctica that appeared to be conducting surveillance operations: Reports came in for approximately 6 weeks detailing "huge spherical craft" above the continent of Antarctica.

These reports came from 5 different sources and described the spheres as being huge, metallic, shiny with one row of portholes going around the sphere. There was speculation that these craft were crew carrying, antigravity *Cosmospheres* now under the control of President Putin and the Russian Federation.

It is feasible that they provided intelligence used by the chevron shaped spacecraft that intercepted and turned back the larger teardrop shaped craft attempting to leave Earth with their likely New World Order global elite passengers. (16)

The SSP Alliance were not sure who the chevron shaped spacecraft belonged to, but the speculation was that they were linked to the "Earth Alliance," a consortium of "White Hats" from various national militaries working closely with the BRICS nations. Apparently, certain factions within our Military Industrial Complex (MIC) have overcome Draco/Nazi control and have decided to become the White Hats. These elements have become part of the Earth Alliance and have partnered with the Russians in fighting the Draco controlled Cabal.

According to intel given to Corey Goode, these chevrons shaped craft were apparently created in two places by Boeing and possibly other contractors. This includes the Skunk Works in Palmdale, and another facility at China Lake in the Mojave Desert.

The underground facility at China Lake is big enough that with the right equipment, it looks like a city of 500,000 people at night. These new, stealthy-looking black triangle craft can fly along normal jets and airliners completely unseen and undetected. They apparently have a rounded area at the front that forms a cockpit. Their total size is approximately 90 x 90 x 90 feet. They are powered by a mercury centrifuge that generates antigravity and propulsion. They can turn on a dime at Mach 17 and it doesn't hurt the people inside.

The whistle-blower, Peter David Beter (http://peterdavidbeter.com/) disclosed the existence of the Cosmospheres in his briefings from the 1970s. Peter Beter said that he had personally witnessed spheres of this sort in hangars, dating back to his time in the 1980s. He said they were 70 to 80 feet in diameter. The black circles around the centre appeared to be portholes with glass over them. That way, whoever was inside had line-of-sight visibility. Pete was never given information about whether these were of extra-terrestrial origin or built by someone here. It does appear that they are perhaps an upgraded version of the reverse-engineered "Cosmospheres" the Russians have had since at least the 1970s.

This therefore suggests that the Russians and the MIC have now teamed up against the Draco and the Cabal, which is a very significant development. Now, if only the U.S. State Department would get on board.

According to Goode, the Antarctic battle was not an isolated incident:

Gonzales reported dozens of underground/ocean conflicts that have involved the use of exotic weapons as well as an uptick in the use of weather modification weapons by both the various syndicates and elements of the Earth Alliance.

Recently, Corey Goode was brought up to the usual location for the Blue Avians. He is standing inside a sphere and sees the stars all around him. The floor he is standing on is a series of criscrossing lines of white light in a square grid pattern. The Blue Avians, such as Tear-Eir, stand in front of him. The communication is telepathic and the sights are very interesting.

He sees multiple spheres in our solar system, with a purplish or bluish color. These are normally cloaked and invisible to our own people, as well as the SSP, but within this area he can see everything. They are all arrayed in a vast grid like pattern throughout our solar system.

The first thing Corey noticed was that the spheres were far more active than usual. There were an ongoing series of energetic discharges passing between them. He was told that they were having to work harder to buffer the ever-increasing changes going on in the sun, as we get closer and closer to this event.

Goode stated that this meeting with Raw Tier Eir occurred on June 10, 2016 and was told that incoming cosmic energies are dramatically ramping up energy levels for the sun, solar system and the entire planet. People and groups are being easily triggered, which helps explains the elite exodus to Antarctica and upsurge of violent extremism around the planet.

One visible sign of these epic changes was in what NASA called a "Monster Hole in the Sun" as of May 27, 2016: http://www.express.co.uk/news/science/674234/Is-the-sun-DISINTEGRATING-NASA-spots-monster-hole-open-up-on-ourstar

Then on July 16, 2016 Corey Goode was again picked up by the usual blue orb. He was taken in the usual manner to the same massive sphere in space. There he could see the same scenes as before. But this time, Gonzales was accompanying other members of the Sphere Being Alliance.

Gonzales looked much more relaxed and happy than his usual demeanor, and asked Corey how he had been with a friendly smile. At their last meeting, Gonzales had not been cordial at all.

After a short greeting, he explained that from his point of view and experience of time, he had been away for almost 6 months. He said that the technology from the Mayan breakaway group is simply magical. The Mayan group consist of intraterrestrials living within the Earth. They are very good at healing the human psyche of emotional and negative thought problems. Corey Goode himself had previously benefited from their healing therapy.

Gonzales stated that he now knew himself, and had found a real purpose in life. He said he had tried to find a purpose in the military by dedicating himself to the mission, and doing whatever was necessary to accomplish it. He had convinced himself that he was working for the greater good, and was in the end a "service to others" type of person. He had fallen into the lower-density trap of egotism and of being self-absorbed.

Gonzales said that his major breakthrough in his "healing" occurred when he answered "yes" to the question do you want to know "who you were, who you are and who you will be" when the Mayan group made the offer.

Then Corey Goode said, "Gonzales then got a serious look on his face and apologized for his recent behavior. He wanted to hear me say out loud that I forgave him, so of course I did. He seemed very relieved that I didn't harbor any resentment towards him." So, the relationship between Corey Goode and Gonzales, after Corey Goode's involuntary "outing" of Gonzales and other team members, was finally healed.

Then, Gonzales was taken away by a blue orb and Raw Tier Eir started discussing with Corey Goode what would happen to the Sun in the near future. Corey explains that he was shown a future scene of the sun:

> "And I also saw an image . . . He showed me an image of one of these waves hitting the Earth, and it wasn't what I expected. I expected a wave just to go 'phoof' through the Earth.
>
> This wave came, and it hits the magnetic field of the Earth, and then it wraps around, and then it comes back on itself in the North and South Pole and then entered to the core of the Earth. And then, the energy emanated from the core of the Earth outward.
>
> The energy was white, but there was an interaction around the magnetic field that looked like aurora borealis but all the way around the Earth.
>
> Tear-Eir tells me that right now is a very pivotal time for

our co-creative consciousness, that everyone on Earth, our co-creative consciousness, was deciding what type of an experience this would be, what type of temporal reality we would go into from this event.

And I wasn't really surprised that the mass consciousness was going to work in tandem with this event. I was surprised at how much it was, how incredibly important the mass consciousness is.

And he further stated that every single person on this planet is just as important as the other. There is no one that is any more special in this process, that we are all partaking in it.

And even some of the non-terrestrial groups that are trapped here are having an influence on it. Their mass consciousness . . . Their consciousness is having an influence on what is happening."

Apparently, the collective mass consciousness of the Earth will affect the outcome or "time line" of how this solar event will manifest in our reality. So, stay positive, loving and in harmony with all living systems on the planet if you wish for a good outcome.

Monday, 13 June, 2016 at 15:31, Luxembourg's Government is a few steps closer to starting its asteroid mining venture after signing a Memorandum of Understanding with Planetary Resources.

The agreement, signed with "Société Nationale de Crédit et d'Investissement" (SNCI) and US-based aerospace technology company, Planetary Resources, provides the framework for their cooperation within Luxembourg's SpaceResources.lu initiative.

The project's goals are the exploration and commercial utilization of resources from Near Earth Objects (NEOs), such as asteroids.

Within this partnership, the government is considering a direct capital investment of 25 million euros in the European headquarters established recently by Planetary Resources in the Grand Duchy.

This public equity position will be taken by the SNCI to become a minority shareholder. For its part, Planetary Resources is contributing to the promotion of the local space industry by developing several key activities exclusively in Luxembourg focused on space hardware development, space services, applied research and scientific discovery.

The Government will support these activities by providing funding through R&D grants or other different types of aid available. Luxembourg Deputy Prime Minister and Minister of the Economy, Étienne Schneider, said: "The Government's partnership with Planetary Resources is another ambitious public-private joint venture which demonstrates our strong commitment to support the national space sector by attracting innovative activities in space resource utilization and other related areas." The agreement paves the way to building up in Luxembourg research activities and technological capabilities in the fields of propulsion development, spacecraft launch integration, deep space communications, asteroid science systems, Earth observation product development and mission operations.

In a speech before the European Parliament on June 28, 2016, in an emergency meeting discussing the consequences of the

Brexit vote, the President of the European Commission, Jean-Claude Juncker made the following statement in French:

> "Il faut savoir que ceux qui nous observent de loin sont très inquiets. J'ai vu et entendu et écouté plusieurs des dirigeants d'autres planètes qui sont très inquiets puisqu'ils s'interrogent sur la voie que l'union européenne va poursuivre. Et donc, if faut rassurer, et les européens, et ceux qui nous observent de plus... loin."

This directly translates into English as:

> "It should be known that those who observe us from afar are very worried. I met and heard and listened to several of the leaders from other planets who are very concerned because they question the path the European Union will engage on. And so, a soothing is needed for both the Europeans and those who observe us from ... farther away."

Of course, the authorities quickly tried to say that Junkers meant something else or tried to change the original text of the speech. But, that is verbatim what he actually said. Were these leaders of other planet CEO's in the ICC? Or were they leaders of native populations of these planets? Why was Jean-Claude Juncker so careless to bring up interplanetary leaders – a highly classified subject in a European Parliament meeting?

What is known, is that certain high level leaders in these secret space programs feel that the world's public have a right to know what is actually going on in outer space and the ET presence and interaction with world governments.

An example of this is William Tomkins who had a five-hour

meeting with Admiral Hugh Webster concerning how much formerly classified information he could place in his planned book. *Selected by Extraterrestrials*. Hugh Webster said, "Bill, tell it all. This is most important to our country. Don't leave anything out."

For another example, let us take the case of Randy Cramer, the Mars Defense super soldier ordered to reveal all that he knew to the public by his commanding officer. Randy explains:

"I am till the day I die, a captain of the United States Marine Corps, special section. USMC s.s. is a covert Unacknowledged Special Access Program (or USAP) signed into law as a legal and covert branch of the US military in 1953 by President Dwight D. Eisenhower. USMC s.s. was mandated by President Eisenhower through the USMC s.s. special code to meet the exopolitical questions of EBE's,(Extra Terrestrial Biological entities) and ETV's (Extra Terrestrial Vehicles); to assist in assessing diplomatic opportunities and military threats and advise the MJ-12 committee and special study groups (SSG's) with intelligence from a fully staffed and operational military intelligence machine, and respond to their requests for specially trained military assistance in all matters extra-terrestrial.

When I agreed to speak publicly, my security clearance was raised to a Blue/Gold-13, which has granted me full access to USMC s.s. intelligence files, and weekly briefings by Brigadier General Julian Smythe, personally."

In essence, Brigadier General Julian Smythe, not only allowed Randy Cramer to disclose what he knew but also gave him access to up to date intelligence since he left the Mars Defense

Force. Randy Cramer was also told that they would "watch his back" in case other forces tried to silence him. Randy Cramer claims that the military had a TR3-B triangular antigravity cruiser surveilling him 24/7/365 and if any group looks like they want to "take out" Randy Cramer, they themselves will be taken out. Randy Cramer's informative web site is here: http://www.earthcitizenconsulting.org/

So, it is possible that Jean-Claude Juncker was also allowed by his secret superiors to leak some information to the parliament members. Or - he could just have been careless.

So, it seems that certain individuals high up in these secret programs are gradually letting these secrets out. But, it is still far from official policy and the news media is not yet on board.

A recent launch attempt by Space X to place a Facebook satellite into orbit, ended in failure. A series of video frames taken of the SpaceX Falcon 9 rocket explosion on September 1, 2016, show two UFOs in the vicinity, just before and during the failed pre-launch test.

The explosion started in the upper portion of the rocket just as one UFO passed close by it. The first part of the explosion was a brilliant white, indicating something other than a fuel combustion, which showed up in later video frames as orange colored with black smoke. This sequence makes it highly like that the UFO was somehow the cause of the explosion.

According to Michael Salla, the video frames are hard evidence linking the SpaceX explosion to an alleged space war being fought between rival secret space programs according to whistleblower Corey Goode: http://exopolitics.org/spacex-rocket-explosion-linked-to-secret-space-war/

The SpaceX rocket was carrying a $150 million satellite belonging to Facebook's Mark Zuckerberg who was quick to tweet his disappointment.

Mind Control experts at CIA Headquarters in Quantico, VA met with WWN in New York on Feb. 17, 2012 to discuss – Facebook.

The CIA experts wanted WWN to keep their identities private, but they went on the record to discuss a detailed strategy that Facebook is utilizing (with help from the CIA) to control the minds of Americans, and citizens around the world.

"Facebook is slowing implementing a plan to get Americans and people around the globe to behave in ways that they believe is best for the planet," said one of the sources. The whole idea of "status updates" was to get users involved in a mindless (and supposedly harmless) activity, posting what they were doing at any moment in the day. But, Facebook has been collecting that "data" and using it to create mind control applications that are being placed on Facebook pages."

I can personally attest that I originally signed up for Facebook in 2010 to see what my grandchildren were doing, as I live on the Big Island of Hawaii and they live on the U.S. mainland. Now, I can't believe the amount of time they spend on Facebook. In my opinion, it has already taken over their lives, time and minds with mostly useless trivia. Perhaps this is the reason for the shoot down of the Space X launch vehicle.

Corey Goode had been attending a number of SSP Alliance meetings where he gained more intelligence on the Earth Alliance. Putin is definitely one leader of the Earth Alliance that is opposing the New World Order cabal. The BRICS financial group also wish to end the cabal's monopoly on global finances. This is why the news media controlled by this Cabal has been falsely vilifying Putin, Russia and China to such a large degree.

Goode also delivers an interesting revelation, the FBI is actively working with Putin's Russia to help get the truth out, and this was part of what motivated the FBI Director Comey's letter reopening the investigation into the Hillary Clinton emails. This is a surprising declaration of collusion between Russia and the FBI by Corey Goode.

Two days before the election on November 6, Comey stated that no charges would be brought at this time against Clinton. The FBI knew that they had enough evidence against the Clintons to as the Chief of the NYPD stated, "put them away for life." In my mind, her greatest crime was the destruction of Libya and the attempted destruction of Syria while she was Secretary of State. Bill Clinton's greatest war crime was the destruction of Yugoslavia on his watch. However, the time wasn't right. Obama would have just pardoned them. It would be better to wait until Trump was President. Goode also stated that the Earth Alliance basically supported Trump.

One of the issues facing the Earth Alliance, whose leaders are in negotiations with cabal leaders is how fast or slow disclosure of the Secret Space Program and ET presence will occur. Cabal leaders wish the disclosure to be a slow one, taking 50 years or more to be finalized. This is because, along with this technological disclosure, many crimes against humanity will also be disclosed, like using slave labor on the Mars colonies. Cabal leaders hope to already be dead by the time their crimes are disclosed, to avoid punishment.

In contrast to more classified space programs which Goode and other whistleblowers claim to have worked with that maintain bases on the Moon, Mars, and deep space fleets, the NRO run space program is primarily limited to near Earth orbit missions. Goode said:

There are space stations that are in near-Earth orbit, upper near-Earth orbit, that are a little bit more advanced than the ISS that they're flying to and from.

Goode insists that personnel involved in the NRO-run space program genuinely believe that it uses the most advanced technologies available to the military industrial complex. He claims that the manned orbital space stations manned by the NRO are effectively a cover program for more highly classified programs that he worked on from 1987 to 2007 that maintain off-world bases and interplanetary fleets.

Goode revealed earlier that the more advanced and highly classified space program called Solar Warden, which maintains interplanetary fleets for solar system wide duties, was set up by the U.S. Navy. In contrast, the less advanced manned orbital spy platforms used by the NRO, were set up by the USAF.

This makes a lot of institutional sense since the USAF is the newest among the U.S. military services. It would be the Navy that would have the institutional history, infrastructure and clout to run the most advanced space program – Solar Warden.

Goode claims that in recent secretive meetings he attended comprising members of a Secret Space Program [SSP] Alliance, there is a plan for the NRO and USAF Space Command to reveal the existence of manned orbital space platforms that are capable of maintaining hundreds of personnel in space.

If they were to do some sort of disclosure and say, this is our secret space program, as a way to trick us and show these craft, these space planes and these space stations, yeah. We would think they're pretty advanced. But they're probably up to 50 years more advanced in the International Space Station.

The goal of the partial disclosure plan is to distract the public from the more damaging revelations by Goode and other space program whistleblowers about off-planet bases, space fleets, and trade with extraterrestrial races. The public will be told that all UFOs are in fact spacecraft belonging to the NRO and USAF Space Command.

With the recent U.S. election of Donald Trump, faster disclosure of these things should take place. Hillary Clinton would have tried to allow a much more restricted disclosure process as she was part of the cabal. She herself could be facing arrest during a Trump Presidency.

Donald Trump could become (or perhaps already is) another Earth Alliance leader, working in cooperation with Russia on things of mutual benefit to both Russia and the U.S. like taking out ISIS, insuring the rule of law on a global scale, and putting an end to the U.S. illegal regime change policies.

The New World Order cabal will try everything in their power to counter all this and among other things, will try to place cabal agents into the inexperienced Trump's cabinet. Now that Donald Trump has been inaugurated President for a week, he has already caused a lot of disruption with his executive orders. Some of it good, like the cancelling of the TPP treaty which would have given corporations even more power than governments. Some of it ill advised, like limiting travel to the U.S. from certain targeted countries, which is against established immigration law.

In Trump's haste to fulfill his campaign promises, it would seem that he requires better legal advice. On the positive side he has already contacted Vladimir Putin by phone and made arraignments to jointly fight terrorism – something that the Obama Administration refused to do. Only time will tell how the Trump presidency will play out.

Conclusions

As more insiders, who have been involved in the Secret Space Program come forward, we are able to get a more detailed picture of just what is really going on in outer space, on Earth and in hidden underground military installations.

We are also being exposed to history formerly hidden from us. How many people were taught in their history class what really happened in the 20th century? Were we taught that the Nazis never surrendered at the end of World War II? Only the appointed head of the German military did. Or, that they moved their best men and technology to secret bases in South America and Antarctica before the war ended?

Or, that with help from fascists in U.S. Industry, Jesuit secret societies like Skull and Bones, and Families like the Rockefellers and the Bushes that the Nazis gained control of the U.S. Military Industrial Complex, infiltrated and took over the CIA and ran NASA? Or, that, in effect that the Nazis won a silent victory over the U.S. by the 1960s? Were we informed by our corporate controlled news media that there was a secret space program, or that the Interplanetary Corporate Conglomerate (ICC) had been formed, or that Mars was being secretly being colonized?

The facts are that a secret breakaway civilization, with their own governing and sources of financing, exist within the United States. This civilization is fascist in nature and operates unlawfully outside of the U.S Constitutional government. And, it is controlled by the same Jesuit/Illuminati organizations that brought the Nazis into power in the 1930s.

This civilization is corporate based and operates a parallel secret space program to the Navy's Solar Warden program outlined by William Tomkins and Corey Goode. The ICC that runs this parallel program is formed from other companies like, Northrop Grumman, Lockheed Martin, Boeing Aerospace, SIAC, Bechtel, and others, but has moved beyond them by colonizing our solar system and beyond.

These companies have Unacknowledged Special Access Projects, on planet and off, that are so secret that even the U.S. President, or the head of military intelligence at the Pentagon don't possess a "need to know" about them.

Also, we are getting a better idea of the advanced technology that has been purposely withheld from our society so that we would keep using cars powered by obsolete internal combustion technology, that hasn't changed much in a century, while polluting our planet.

While in college in the 1980s, I myself knew that the science taught us at the universities had large omissions and false conclusions. For example, nowhere was it taught that Thomas Townsend Brown had experimentally demonstrated the connection between gravity and high voltage electric fields, and had patented antigravity, and free energy devices based on his experiments in the 1920s. Also, we were taught that Einstein was unable to create a theory that unified gravity with electromagnetism.

Thomas Bearden points out that Maxwell's original theory did just that. But, his original theory wasn't being taught in the universities.

This is so that the oil companies will continue to profit while global warming is causing ever more severe weather patterns. Near sighted greed blinds the CEOs from farsighted consequences.

In the area of spirituality, the Sphere Being Alliance has a message for humankind which is quite similar to the one Jesus Christ brought 2,000 years ago:

> "Every day focus on becoming more service to others oriented. Focus on being more loving and focus on raising your vibrational and consciousness level and to learn to forgive yourself and others (thus releasing Karma). This will change the vibration of the planet, elevate the shared consciousness of humanity and change humanity one person at a time (even if that one person is yourself)."

Corey Goode also added:

> "They say to treat your body as a temple and change over to a "Higher Vibrational Diet" to aid in the other changes. This sounds to many like a "Hippy Love and Peace" message that will not make a difference. I assure you the "Path" they lay out in "Their Message" is a difficult one. Even on the unlikely chance that these technologies stay "Suppressed", imagine what a world we would live in if everyone made these changes to their selves?
>
> The Blue Avians also gave a warning with this message. They had tried to deliver this message "Three Other

Times" and it had been distorted by humanity. They made it "very clear" that this information was NOT to become a "Cult or Religious Movement" nor was I to put myself (my "ego") before the "Message" or elevate myself to a "Guru" status. Anyone who does so should be avoided and held accountable, including myself."

At a time of great change, where global elites are finding refugee in remote locations amidst uncertainty over what the future holds for them, and extraterrestrials debate over whether to openly make contact, the choice is ours over what the future holds for us.

We must remember the power of love and forgiveness in opening up positive possibilities for us all, including that of a peaceful full disclosure of the SSPs and extraterrestrial alliances that can greatly expand and enrich human destiny.

Also, we see in the chapter on other science, that there is a new plasma science being developed by Keshe and the Keshe Foundation which already has a commercial free energy generator on the market. To Keshe's credit, he is offering the knowledge of his technology to the world without proprietary secrecy, free of charge.

His strategy is quite sound. By creating international schools and teachers of his technology, he knows that the technology cannot be suppressed - as has happened to so many inventors working alone in the past.

One product under development is the *Oasis Unit*. This unit will not only provide energy to cook food and clean potable water but also actually create the food to cook! China has shown a large interest in the *Oasis Unit*. The *Oasis Unit* is a spinoff of the

Keshe Foundation Space Ship Institute (KFSSI). They discovered that the plasma generator that generated antigravity with certain additions could also produce substances needed for extended space travel. Substances like; oxygen, water, and complex protein molecules that could actually be used for food. This product could also be a large help in poor countries or areas that face lack of potable water and famine.

Consequently, the Keshe Foundation is definitely providing solutions to the world's problems in many areas. So, would the suppressed technology used in the Secret Space Programs - if put to use in serving humanity instead of being kept secret.

According to Corey Goode, there are quite a lot of classified technologies that are in use in these Secret Space Programs that are being suppressed and could completely change the nature and quality of life of every human being here on Earth. The "Free Energy" technologies would end the need of the "Current Oil/Petro Energy Companies", The "Frequency and Light Healing" technologies would end the "Current Pharmaceutical Corporations", the "Neurological Interface" technologies would end the need for "Large Education Institutions" and the "Food Replication" technologies and "Environmental Purification and Restoration" technologies would end poverty, starvation and begin to reverse the ecological damage humanity has done to the Earth.

As you can imagine, the real threat to disclosure is not that humanity cannot handle the truth or will be able to reconcile "Cosmic Life" with their religious beliefs. The real reason is that these technologies would threaten to collapse the world economies and make the present economic system of no use anymore. Many corporations and countries that depend on selling oil products will be forced to diversify into other means of revenue creation.

I believe that this would be a gradual process as all the people that own fuel guzzling cars cannot immediately purchase a new free energy car or antigravity car. Also, not everyone can afford to buy a free energy generator, or even an independent of the grid solar system for that matter, for their home right away. For these reasons, it could take a decade or two to gradually phase out present systems. Even if petroleum based fuel eventually became obsolete, petroleum would still be used for plastics and other products.

It will mean the eventual loss of control of the 0.01% elite over the masses and a complete paradigm change. In short it means freedom! Real freedom by replacing many of the means of economic bondage for the first time in humanity's known recorded history.

There has been a Stealth Civil War going on among various SSP Alliance groups for a while to achieve this result. Relatively recently, the SSP Alliance was joined by a group of beings that no one had encountered before. They are a 6-9th density group of beings that have been referred to as The Sphere Being Alliance. These new Sphere Beings have since not only created an energetic blockade around Earth but have also done so around our entire solar system.

And, according to Corey Goode, one of the most knowledgeable of the Secret Space program insiders, the real data dumps and full disclosures are waiting for the proper circumstances. The Cabal, meanwhile, prefers partial disclosure presented as fiction, like the new *X-files*.

This will give them more time to escape the inevitable of being held accountable for their various crimes against humanity.

The SSP Alliance feels that full disclosure could occur after a climatic event, like a financial collapse, when the masses will be more in a frame of mind to receive the information. Consequently, this full disclosure is yet to come.

I prefer not to wait for a financial collapse. In fact, I don't even want a financial collapse. Things are too difficult for the working class people already. One fifth of our nation's young people have to live with their parents or grandparents because they cannot afford to rent or buy their own home. And, the homeless problem just keeps increasing.

I think that we are ready for full disclosure of the suppressed technology that would bring more independence and prosperity for everyone on the planet - not sometime in the future - but RIGHT NOW!

END

Other Books by this Author:

1. *The Adventurer: The Autobiography of Herbert Grove Dorsey III* https://www.amazon.com/Adventurer-Autobiography-Herbert-Grove-Dorsey/dp/1432771868/

2. *The Secret History of the New World Order* https://www.amazon.com/Secret-History-New-World-Order/dp/147873521X/

3. *Secret Science and the Secret Space Program* https://www.amazon.com/Secret-Science-Space-Program/dp/057815238X/

4. *CIA: Crime Incorporated of America* https://www.amazon.com/CIA-Incorporated-Herbert-Dorsey-III/dp/1478757930/

5. *The Covert Colonization of Our Solar System* https://www.amazon.com/Covert-Colonization-Our-Solar-System/dp/1478768835/

Bibliography

1. *Secret Science and the Secret Space program* by Herbert Dorsey

2. *The Covert Colonization of Our Solar System* by Herbert Dorsey

3. http://www.ufocrashbook.com/eisenhower.html

4. *Selected by Extraterrestrials* by William Tompkins

5. *Energy from the Vacuum* by Thomas Bearden

6. *http://spherebeingalliance.com/blog/transcript-cosmic-disclosure-empaths-and-extraterrestrials-with-clifford-stone.html*

7. http://www.ufodigest.com/mystery.html

8. *The Montauk Project: Experiment in Time* by Peter Moon and Preston B. Nichols

9. http://www.digitalmontauk.com/

10. http://spherebeingalliance.com/blog/corey-goode-intel-update-part-1-aug-2016.html

11. *https://www.amazon.com/ALTERNATIVE-Sci-Fi-Classic-Republished-Material-ebook/dp/B0182YIKNE*

12. *The Ascension Mystery* by David Wilcox

13. *http://www.themarsrecords.com/wp/category/the-mars-records/*

14. *CIA: Crime Incorporated of America* by Herbert Dorsey

15. *http://spherebeingalliance.com/blog/transcript-cosmic-disclosure-from-venus-to-antarctica.html*

16. *http://exopolitics.org/secret-space-programs-battle-over-antarctic-skies-during-global-elite-exodus/*

CPSIA information can be obtained
at www.ICGtesting.com
Printed in the USA
LVOW10s0717260317
528469LV00009B/464/P